FOOD MICROBIOLOGY LABORATORY

CRC Series in
CONTEMPORARY FOOD SCIENCE

Fergus M. Clydesdale, Series Editor
University of Massachusetts, Amherst

Published Titles:

New Food Product Development: From Concept to Marketplace
Gordon W. Fuller

Food Properties Handbook
Shafiur Rahman

Aseptic Processing and Packaging of Foods: Food Industry Perspectives
Jarius David, V. R. Carlson, and Ralph Graves

Handbook of Food Spoilage Yeasts
Tibor Deak and Larry R. Beauchat

Getting the Most Out of Your Consultant: A Guide to Selection Through Implementation
Gordon W. Fuller

Food Emulsions: Principles, Practice, and Techniques
David Julian McClements

Antioxidant Status, Diet, Nutrition, and Health
Andreas M. Papas

Food Shelf Life Stability
N.A. Michael Eskin and David S. Robinson

Bread Staling
Pavinee Chinachoti and Yael Vodovotz

Food Consumers and the Food Industry
Gordon W. Fuller

Interdisciplinary Food Safety Research
Neal M. Hooker and Elsa A. Murano

Automation for Food Engineering: Food Quality Quantization and Process Control
Yanbo Huang, A. Dale Whittaker, and Ronald E. Lacey

Introduction to Food Biotechnology
Perry Johnson-Green

The Food Chemistry Laboratory: A Manual for Experimental Foods, Dietetics, and Food Scientists, Second Edition
Connie M. Weaver and James R. Daniel

Modeling Microbial Responses in Food
Robin C. McKellar and Xuewen Lu

CRC Series in
CONTEMPORARY FOOD SCIENCE

FOOD MICROBIOLOGY LABORATORY

Lynne McLandsborough

CRC PRESS

Boca Raton   London   New York   Washington, D.C.

Library of Congress Cataloging-in-Publication Data

McLandsborough, Lynne Ann.
    Food microbiology laboratory / Lynne A. McLandsborough.
        p. cm. — (CRC series in contenporary food science)
    Includes bibliographical references and index.
    ISBN 0-8493-1267-1 (alk. paper)
    1. Food—Microbiology—Laboratory manuals. I. Title. II. Series.

QR115.M397 2003
664'.001'579—dc21

2003046140

Visit the CRC Press Web site at www.crcpress.com

© 2005 by CRC Press LLC

No claim to original U.S. Government works
International Standard Book Number 0-8493-1267-1
Library of Congress Card Number 2003046140
Printed in the United States of America   4 5 6 7 8 9 0
Printed on acid-free paper

Preface

Microbiology is a laboratory science. As an undergraduate, I was a good general science student and did well in my classes, regardless of the subject matter. However, I thought that laboratories were tedious exercises that rarely enhanced the information being taught in lecture, until I took my first course in microbiology. I have been a "student" of microbiology for the past 20 years, and I still believe that to understand basic microbiology (and food microbiology), one needs some experience in the laboratory. For this reason, I believe that lectures cannot be separated from a concurrent food microbiology laboratory, and the two should complement each other. These laboratory exercises evolved over 8 years of teaching. Students using these exercises range from food science seniors who have taken an introductory microbiology course to dietetics majors who have no background in microbiology. For this reason, the first laboratories cover basic techniques in depth, and schematics of dilution schemes are included for all exercises. In addition, all parameters and dilutions presented in this text have been optimized to ensure the success of each exercise, because time and money constraints prevent classroom laboratory instructors from allowing students to learn through experimental failures.

This text is not intended to be a comprehensive guide to all techniques and the detection of all organisms from foods. Instead, presented are 18 exercises that cover the basic concepts of food microbiology, including variations of detection and enumeration assays. I encourage instructors to use these exercises as the backbone of the laboratory session and incorporate other exercises or test kits to reflect the emphasis of your classes. Typically, I use eight to nine of these laboratories and add in commercial rapid test kits, depending upon my budget each semester. For example, in Laboratory 3, we will often use a commercial MPN for coliforms (SimPlates™, BioControl Systems, Inc., Bellevue, Washington) in addition to the traditional three-tube most probable number (MPN). A commercial enzyme-linked immunosorbent assay (ELISA) kit or polymerase chain reaction (PCR) will often be incorporated in the pathogen labs to build on difficult topics for students to conceptualize. A useful source for information about rapidly changing test kits and whether or not they have Association of Analytical Chemists (AOAC) verification can be found at http://www.aoac.org/testkits/microbiology-kits.htm.

Acknowledgments

I greatly appreciate everyone who has given me invaluable help and assistance in assembling these laboratory exercises. I would like to thank Ron Labbe and Robert Levin for their insights and experience. My first teaching assistant William K. Shaw, Jr., developed Laboratories 7 and 15 and helped in every aspect of the evolution of this text. Emmanouil Apostilidis and Chris Kosteck optimized the parameters used in Laboratories 14 and 18, respectively. John Wood, a highly creative undergraduate in our department, drew the fungal illustrations in Laboratory 2. In addition, my deepest thanks go to those who took the time to review this manuscript: William K. Shaw, Jr., Caroline Cronin, and Marcus Teixeira. Finally, I want to thank my husband Edward, son Aaron, and daughter Sophia for their love, support, and patience during this project.

Lynne A. McLandsborough
University of Massachusetts, Amherst

The Author

Lynne A. McLandsborough, Ph.D., is an associate professor in the department of food science, University of Massachusetts, Amherst, MA.

Dr. McLandsborough received her B.A. degree in microbiology from Miami University (Ohio) in 1986. She received her M.S. and Ph.D. degrees in food science from the University of Minnesota in 1989 and 1993, respectively. She held a postdoctoral fellowship in the department of microbiology at the University of Minnesota before joining the department of food science at the University of Massachusetts in 1995. Her research interests include the mechanisms of microbial adhesion, the ecology of biofilm formation, and the methods of bacterial removal from processing surfaces.

Dr. McLandsborough is a member of the American Society of Microbiology, the Institute of Food Technologists, the International Association for Food Protection, and the New England Society of Industrial Microbiology. She is currently an associate editor of the *Journal of the Science of Food and Agriculture* and is on the editorial board for *Food Biotechnology* and has served on numerous federal grant review panels. She teaches food microbiology (for food science majors) and hygienic handling of foods (to dietetic majors). In recognition of her teaching efforts, she recently received the University of Massachusetts College of Food and Natural Resources Outstanding Teaching Award.

LABORATORY SAFETY

This is the most important information in this text. It is crucial for students to work safely in a microbiology laboratory. In this class, you will be isolating a variety of organisms from foods. The foods will be a mixture of microorganisms: some of these may be nonpathogenic, while other isolates may be pathogenic. Consequently, it is very important that all samples be treated as though they contain pathogens.

STANDARD PRACTICES FOR MICROBIOLOGY LABS

Safety Equipment

1. **Lab coats and closed toe shoes** should be worn at all times. Lab coats may be supplied as part of the course, or purchase may be required. If purchased, lab coats should be kept in the teaching laboratory and autoclaved before students are allowed to take them home at the end of the semester.

2. **Eye protection** should be worn when working with cultures that may contain high levels of human pathogens. Eye protection should be worn at all times for contact lens wearers. For sanitary purposes, every student should purchase his or her own eye protection.

3. **Nonlatex gloves** should always be available. If the skin on the hands is broken, always apply a bandage and wear gloves. For any work with a pathogen or enrichment for a pathogen, gloves should be worn at all times. After wearing, gloves should be disposed of as biohazard waste, and hands should be washed.

Standard Practices

1. **Arrive to lab prepared.** Read and study each lab exercise before coming to class to make yourself aware of potential hazards.

2. **Do not eat, drink, apply cosmetics, or handle contact lenses in the teaching laboratory.** The food samples used in class are not for consumption.

3. **Wash hands frequently.** People should wash their hands after handling any cultures, after removing gloves, and before leaving the laboratory for the day.

4. **Sanitize work area.** Benchtops should be washed down with sanitizer before starting work and before leaving for the day.

5. **Be aware of your laboratory environment.** Take notice of where fire extinguishers are stored, where the eye wash station is located, and where the nearest phone is located in case of emergency.

6. **Use open flames safely.** Gas burners should be turned off when not in use and definitely before leaving the laboratory. Tie long hair back so it does not get into the Bunsen burner flame. Keep ethanol at least 18 in. from open flame.

7. **Keep your work area organized.** Bring only your lab notebook to the lab bench. Coats, backpacks, and purses should be kept in a designated area away from workbenches. Work in

an organized fashion with your partner, because multiple people working in a small space can lead to hazardous mistakes.

8. **Notify the instructor immediately if you or another student are injured in any way.**

9. **Notify your instructor of any spills that occur.** If any portion of a culture or contaminated equipment comes into contact with the lab bench, floor, or equipment, the area should immediately be covered with a paper towel and flooded with sanitizer. Notify the instructor or teaching assistant after applying the sanitizer. If spills occur on gloves, notify the instructor immediately. Gloves should be removed and placed in a biohazard disposal area, and hands should be washed with soap and hot water for at least 30 sec.

10. **Notify your instructor of any broken equipment or unsafe practices.** Notify your instructor if a piece of equipment is broken to avoid potential safety problems. If you feel that your partner or others in the laboratory are working in a potentially dangerous manner, notify the instructor.

11. **Work slowly and carefully.** Rushing is the cause of most lab accidents.

12. **Relax and have fun.** The act of performing an experiment is a small portion of the work that your instructor spends on these laboratories. The majority of time and work for these exercises is spent planning, preparing supplies, and cleaning up. Be appreciative of your instructor for the time and energy it takes to have everything ready so that students can walk in and have the supplies ready to do the fun portion of each laboratory session.

Table of Contents

LABORATORY 1

FISH MICROFLORA: BASIC MICROBIOLOGICAL TECHNIQUES AND STANDARD PLATE COUNTS

■ I. OBJECTIVES

- To master dilutions, pour plates, and spread plates.
- Use plate counting guidelines to calculate CFU/g (colony-forming unit per gram).
- Learn to streak plate for purified cultures.

■ II. BACKGROUND

Sampling and Preparing Food for Bacteriological Analysis

One important aspect of food microbiology is that bacteria are usually heterogeneously distributed within food products. On commodities, such as fruit, vegetables, meats, and fish, the bacterial load will likely be higher on the surface when compared to the interior of the item. In addition, distribution can vary within a given product. For example, within different portions of a fish fillet, bacteria are usually unevenly distributed, with higher numbers around the fin and gut areas. This also holds true for processed foods. Often, bacteria are not distributed homogeneously through an entire lot of food. For example, if a standard plate count (SPC) was performed on a sample from a single unit of a 1000-unit lot, one cannot know if the results are representative of the entire lot or if they are exceptions. Therefore, analyses of a greater number of samples will give a broader understanding of the food product's microbial quality. In addition, greater sample numbers increase the probability of finding a product containing high microbial numbers or even a pathogen within the sampled product. However, lab supplies, personnel, and product costs must be considered in the cost-effective operation of the laboratory prior to performing an analysis on the number of samples needed for each analysis. Statistical sampling plans can assist in determining the most appropriate sample number to assure a given level of risk in a food product.[1-3]

Care must be taken in collecting food samples and transporting them to the laboratory for analysis. It is ideal to submit samples to the laboratory in unopened containers. Otherwise, leakproof containers and sterile stainless steel utensils should be used for sampling and transport. Frozen samples should remain frozen during transport; refrigerated samples should not be frozen, but should be kept between 0 to 4°C during transport. All samples should be examined within 24 h of reaching the laboratory. Frozen samples should be stored frozen, and perishable refrigerated items should be stored at 0 to 4°C.

Once in the laboratory, samples of 25 to 50 g are typically used for analysis. Prior to opening a sample, the surface of the food container should be sanitized with 70% ethanol in order to reduce the incidence of unintentional contamination. Liquids should be mixed by inversion before sampling with a sterile pipette. Solids must be sampled using sterile utensils (knives, spoons, cork borers) and must be weighed prior to blending with diluents.

Diluents often used in food microbiology include Butterfield's phosphate-buffered dilution water (0.6 mM KH_2PO_4, pH 7.2), serological saline (0.85% w/v NaCl), and peptone water (0.1% w/v peptone), or in Europe they use a combination peptone saline water (0.85% NaCl and 0.1% peptone).[1,2,4] The reason for the addition of peptone and salt is to maintain the osmotic stability of diluted cells. This works well as long as the diluted cells do not sit for longer than 30 min. After 30 min, bacteria can grow in peptone water, and an extended time in saline can accelerate cell death. For this reason, whenever performing bacterial enumeration, it is best to dilute and plate within 30 min of homogenizing the product.

Analyses of solid food products are usually performed after blending. Blending can be performed by weighing a sample into a sterile blender cup or into a sterile plastic bag. Diluent is added (usually nine times the sample weight), and food is either blended using a blender (Waring® or other heavy-duty blender) or masticated using a laboratory paddle blender (Stomacher® or other brand of paddle blender).

Dilution Basics

To obtain accurate quantitative analyses of cell numbers, petri dishes should have relatively diluted bacterial samples (25 to 250 CFU/plate). At the time of plating, you will never know the exact number of cells in any solution (although you may have an educated idea of the levels you expect to find). For this reason, a series of dilutions are always plated with the purpose of finding at least one dilution with plates in the countable range.

Serial dilutions are a series of dilutions. In bacteriological work, dilutions are usually performed in series of 1/10 or 1/100 dilutions. A series is used because it allows us to take samples and analyze at different concentrations. Before we look at a series, let us review simple dilutions. A 1/10 dilution consists of a 1 ml volume of sample added to 9 ml volume of diluent, 11 ml volume of sample to 99 ml volume of diluent, or 25 g of food sample to 225 ml diluent (see Equations 1.1, 1.2, and 1.3, respectively). When calculating the total dilution, the sample volume is added to the diluent volume (also called a blank). Also note that when solid foods are used, it is assumed that 1 g of food is equivalent to a volume of 1 ml. In order to simplify a dilution series, scientific notation is used, and a 1/10 dilution can be expressed as 0.1, 1×10^{-1} or simply 10^{-1}:

$$\frac{1 \text{ ml}}{9 \text{ ml} + 1 \text{ ml}} = \frac{1 \text{ ml}}{10 \text{ ml}} = 1 \times 10^{-1} = 10^{-1} \tag{1.1}$$

$$\frac{11 \text{ ml}}{99 \text{ ml} + 11 \text{ ml}} = \frac{11 \text{ ml}}{110 \text{ ml}} = \frac{1 \text{ ml}}{10 \text{ ml}} = 0.1 = 1 \times 10^{-1} = 10^{-1} \tag{1.2}$$

$$\frac{25 \text{ g}}{225 \text{ ml} + 25 \text{ g}} = \frac{25}{250} = \frac{1}{10} = 0.1 = 1 \times 10^{-1} = 10^{-1} \tag{1.3}$$

The other commonly used dilutions in food microbiology are 1/100 dilutions. These are performed the same way. They can be 1 ml added to 99 ml or 0.1 ml added to 9.9 ml (Equation 1.4).

$$\frac{1 \text{ ml}}{99 \text{ ml} + 1 \text{ ml}} = \frac{1 \text{ ml}}{100 \text{ ml}} = 0.01 \text{ ml} = 1 \times 10^{-2} = 10^{-2} \tag{1.4}$$

Serial dilutions are performed with a series of dilutions. Figure 1.1 shows a basic dilution scheme using 9 ml blanks. It is prepared as follows:

1. The initial 1/10 dilution (1 ml into 9 ml) is performed.

2. This is mixed using a vortex mixer.

3. A volume (1 ml) is taken, mixed, and added to the next tube for the second 1/10 dilution. These dilutions are additive; therefore, the second tube in the series of two 1/10 dilutions has the final dilution of 10^{-2} ($1/10 \times 1/10 = 1/100$ or 10^{-2}).

4. The 10^{-2} dilution is mixed, and a 1 ml sample is removed and added to the 9 ml tube. As before, the dilutions are additive; therefore, this dilution is 10^{-3} ($1/10 \times 1/10 \times 1/10 = 1/1000$ or 10^{-3}). You can also think about it as $10^{-2} \times 10^{-1} = 10^{-3}$. This continues, as you can see in Figure 1.1.

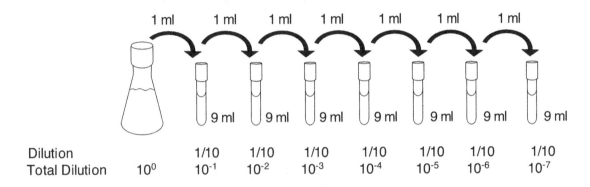

Figure 1.1 Simple serial dilution series using 9 ml blanks.

Theoretically, you could make these dilutions any way you want. You could add 1 ml into 999 ml to get a 10^{-3} dilution, but this would be impractical. For example, if you performed a single dilution to obtain a 10^{-6}, you would need to add 1 ml into 1000 liters, or 1 μl into 1 liter. You can see why serial dilutions are simple, use relatively small volumes, and allow solutions to be diluted with minimal error.

The objective of a plate count is to determine the number of organisms in the food at the time of analysis. A dilution scheme with plating can be seen in Figure 1.2. Cell numbers are always expressed as CFUs, because it is not known if the counted colony grew from a single cell or a clump of cells. When performing a SPC, the number of organisms that are in the food sample is unknown. It is the responsibility of the technician performing the test to design it with a wide enough range of dilutions to ensure that after incubation, some dilutions fall within the countable range (25 to 250 colonies/plate). From this data, the initial cell number (expressed as CFU/g for solid foods or CFU/ml for liquid foods) is calculated. If you plate 1 ml from your dilution, the final dilution you plate is the same as the dilution in the dilution bank. If you plate a 0.1 ml volume (as with spread plates [Figure 1.2]), this is considered an additional 1/10 dilution. If you are plating 0.1 ml from a 10^{-6} dilution, the "dilution as plated" is $0.1 \times 10^{-6} = 10^{-7}$. To eliminate errors, plates should always be labeled with "dilution as plated."

Standard Plate Counts

A SPC (or aerobic plate count [APC]) is used to determine the level of microorganisms in a food product or an ingredient. These data are often used as indicators of food quality or predictors for the shelf life of a product. The SPCs use media without any selective or differential additives. Sometimes a SPC may be referred to as a "total plate count," which is a misleading name. It is important to remember that colonies, which grow during the incubation period, do not represent the entire microbial population of the food product. The colonies counted during a SPC only represent the organisms

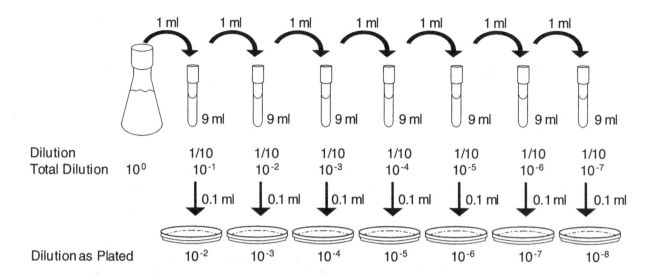

Figure 1.2 Simple dilution scheme with plating.

that could grow at the growth conditions we defined (temperature, incubation period, media, and atmosphere). Other populations of organisms that can only grow at higher or lower temperatures, grow very slowly, require additional nutritional components, or need a specialized atmosphere (such as a reduction in O_2 or increase in CO_2 or N_2) and will not be part of the SPC. As with all culture-based tests in microbiology, results obtained are influenced by the "window" of our culture conditions. If growth conditions are changed, the organisms observed to grow may or may not be different. Therefore, it is important to use standardized conditions in order to compare data from the same laboratory over time or data from different laboratories.

SPCs can be performed using pour plate or spread plate techniques. In pour plates, 1 or 0.1 ml samples from dilutions are pipetted into sterile empty petri dishes. Tempered (45°C) agar (approximately 20 to 25 ml) is then added to each plate, and plates are mixed by swirling on a flat surface.[1,4] In the spread plate method, small volumes of diluted sample (0.1 ml) are spread onto solidified agar plates with a sterile bent glass rod. Table 1.1 outlines the major differences between these two methods. Regardless of which method is used, each dilution should be plated in duplicate or triplicate, inverted, and incubated at 35°C for 48 h. (Dairy products should be incubated at 32°C for 48 h.)

Guidelines for Colony Counts

After incubation, plates should be counted according to the following guidelines adapted from *The Compendium of Methods for the Microbiological Examination of Foods*.[1] The purpose of these guidelines is to assure reproducibility between different researchers and laboratories.

> **Guideline 1: Any colonies that are physically touching are counted as one.** Colonies of some bacteria may be irregular in shape, and therefore, it is difficult to tell how many cells were derived from the "touching" colonies. This counting may be interpreted differently by each researcher. To standardize counting, any colonies that are touching are counted as one. Make sure that all colonies are counted, including any pinpoint colonies. A colony counter with back illumination can aid in viewing small colonies.
>
> **Guideline 2: Count plates from all dilutions containing 25 to 250 colonies**. Average the counts from replicate plates of the same dilution, and multiply the average number by the reciprocal of the dilution used. (This will give the CFU/ml or g of the initial product.) Report the dilutions used, the count observed for each dilution, and the calculated CFU/ml or g of the initial product.

TABLE 1.1

Comparison of Pour Plates vs. Spread Plates

Pour Plate	Spread Plate
Growth of organisms that need lower oxygen tension (such as microaerophilic organisms or injured cells) may be favored	Growth of injured or microaerophilic cells may be reduced, because the agar surface has a higher oxygen tension
Volume plated (0.1 or 1 ml samples) can be varied; larger volumes (1 ml) may be more accurate in detecting low cell numbers	Only small volumes (0.1 to 0.5 ml) can be plated; this may be less accurate for detecting low levels of bacteria
Colony morphology is not usually observed, because most colonies are imbedded in agar	Colony morphology and pigmentation are easily observed
Imbedded colonies do not tend to spread throughout the plate	"Spreaders" can be a problem
Warm agar use may inhibit the growth of highly heat-sensitive organisms	In some cases, higher counts may occur because organisms are not exposed to warm agar
Agar must be steamed and tempered (45°C) before use	Agar plates can be prepared and dried ahead of time
Agar medium must be translucent	Translucent or opaque media can be used

Source: Adapted from Swanson, K.M.J., Busta, F.F., Peterson, E.H., and Johnson, M.G., in *Compendium for the Microbiological Examination of Foods*, 3rd ed., American Public Health Association, Washington, D.C., 1992.

Guideline 3: If only one plate of a duplicate pair yields 25 to 250 colonies, count both plates and average the counts. If no other dilution falls within the 25 to 250 colony range, use this for calculating CFU/ml or g for the initial product. Do not use plates with spreaders.

Guideline 4: If consecutive dilutions have 25 to 250 (for example, both 10^{-4} and 10^{-5}), calculate the CFU/g from each dilution and report the average as the CFU/g. But if the higher count is more than twice the lower count, report the lower computed count as CFU/g.

Guideline 5: If all plates have more than 250 colonies, select the most diluted sample and estimate by counting a portion of the plate. Use a colony counter with a guide plate ruled in square centimeters, if available. When there are fewer than 10 colonies/cm, count 12×1 cm^2 squares and calculate the average/cm^2. When there are more than 10 colonies/cm, only 4×1 cm^2 squares need to be counted and used to calculate the average/cm^2. The area of a 15×100 petri plate is approximately 56 cm^2. Multiply the average number/cm^2 by 56 to determine the colonies/plate. Then multiply this number by the reciprocal of the dilution to determine the estimated CFU/g.

Guideline 6: If all plates have fewer than 25 colonies, record the number of colonies on the lowest dilution and report the count as the estimated CFU/g.

Guideline 7: If no colonies are detected on any plates, report the estimated count as less than (<) 1 times the reciprocal of least dilute (or greatest) plated dilution. For example, if you plated 10^{-3} to 10^{-7} dilutions and did not see colonies on any plate, report this as $<1 \times 10^3$ CFU/g.

Guideline 8: Spreaders. There are some bacterial types that just naturally swarm and take over an agar plate. (Members of the *Proteus* genus are notorious for swarming and causing "spreaders.") When spreading colonies are observed, count each spreading colony as one unless it exceeds 50% of the plate area. When the colony exceeds 50% of the plate, report this as a "spreader" (spr). These plates cannot be used in calculations.

Calculation of SPC

One of the most confusing parts of SPCs is the calculation of CFU/g in your original product. Study this example carefully. The data in Table 1.2 represent the counts from an experiment. Two petri dishes were inoculated from each dilution. The only dilution with 25 to 250 colonies was 10^{-4}. To calculate the original CFU/g in your product, you need to take the average of the countable dilution and multiply it by the inverse of your dilution. An easy trick to use with 1/10 or 1/100 dilutions is to calculate the inverse of the dilution by removing the negative sign. For example, the inverse of 1×10^{-4} is 1×10^4. (Please note that this trick only works if the number in front of the exponent is one. For example, the inverse of 5×10^{-2} is 2×10^1, not 5×10^2.)

TABLE 1.2

Example Experimental Data

| Dilution as Plated | CFU Counted/Plate | |
	Plate 1	Plate 2
10^{-3}	TNTC[a]	TNTC
10^{-4}	150	120
10^{-5}	15	10
10^{-6}	1	0

[a] TNTC = Too numerous to count (>250/plate).

Here is how the calculation is performed:

Average CFU/plate $\times$ 1/dilution = CFU/g

Average CFU/plate = 150 + 120/2 = 135 average CFU/plate

CFU/g = 135 Average CFU/plate $\times 10^4$ = 135×10^4 = 1.35×10^6 CFU/g

Finally, in order to avoid a false impression of accuracy, SPCs should only be recorded to the first two significant digits. This is done by rounding down if the third digit is one through four or up when the third digit is six through nine. When the third digit is five, it is rounded up when the second digit is odd and down when the second digit is even. Therefore, the reported CFU/g for our example is as follows:

CFU/g = 1.4×10^6 CFU/g

Mechanical Dilution: Streak Plate Method

The purpose of streaking a culture on a plate is to dilute the culture enough to get isolated colonies. Isolated colonies are needed to define different colonial morphologies and detect different biochemical characteristics. In addition, streak plating is performed to purify bacterial cultures before further analysis is performed. Because bacteria often exist as clumps or chains, stringent colony purification involves streaking isolated colonies at least twice: first to isolate a colony, then a second time to assure that the clump or chain that started the initial colony was homogeneous.

There are many different streaking patterns for isolating colonies. As long as a researcher achieves the main objective (isolated colonies), no methodology is wrong. However, for most people, streaking for isolation is a technique that takes practice. Two of the methods in common use are discussed later in this chapter.

■ III. METHODS

Class 1: Enumeration and Basic Characterization of Fresh Fish Fillet Microflora

Sample Preparation

1. Aseptically measure (using a sterile utensil) 25 g fresh fillet and place into a sterile Stomacher bag.
2. Add 225 ml sterile peptone water (0.1% peptone in water).
3. Place bag into "Stomacher 400" for 2 min.
4. Use a beaker to hold bag.
5. Prepare a series of 1/10 dilutions, as shown in Figure 1.3A and Figure 1.3B. You will plate 10^{-4} through 10^{-8} dilutions. (Remember that our blending was our initial 1/10 dilution.)
6. Open the bag, and withdraw a 1 ml sample. Avoid sampling foam. This initial sample can be tricky, because larger portions of homogenized fish can clog your pipette. One solution for avoiding pipette clogging is to use wide bore pipettes or break the tip off of a disposable 10 ml plastic pipette.
7. Between each dilution, mix samples by shaking all dilutions 25 times in a 30 cm (1 ft) arc within 7 sec, with caps screwed on tightly. If the dilution bottle was standing for more than 3 min before plating, shake the dilution again before transferring or plating. If using test tubes (9 ml dilutions), vortex on full speed for approximately 7 sec for adequate mixing.

Pour Plate Method
Procedure

1. Label the bottoms of empty plates (names, pours, and dilutions as plated). You will be plating each dilution in duplicate.
2. Pipette 1 ml of the appropriate dilution (Figure 1.3A) in each plate.
3. Once all samples are placed in the petri dishes, get a bottle of tempered (45 to 47°C) liquid plate count agar from the water bath. Each bottle has approximately 100 ml of agar, which will be enough for four pour plates (approximately 25 ml agar per plate). You will need two bottles, but get one at a time, because agar will start to solidify at 40 to 43°C. If bottles are left out of the water bath too long, you will get solid agar clumps in your plates, making it much harder to count colonies later.
4. Gently pour in tempered agar while swirling the plate gently. Immediately after everything is poured, go back and swirl the plates gently on top of the bench to assure even mixing. Carefully moving the plates in gentle, slow, figure-eight motions for about 5 sec should be sufficient. With practice, two to four plates can be swirled at the same time. However, be careful, if you mix too roughly, you can splash agar onto the top of the dish, which can reduce the accuracy of the results.
5. As you pour, leave 1 to 3 ml of agar in the bottle. Pour this remaining agar into a blank plate (no sample). Do not worry if there are just a few puddles sitting in a plate. Incubate this plate along with your samples as your agar control. If you obtain growth in this, you will know that your agar was contaminated, and the results should be discarded.
6. After the agar bottles are empty, rinse them with water and place on a designated cart. It is important that molten agar is never put down the sink (even though it keeps the plumbers in business). If there is a substantial amount of sterile agar remaining, pour it into a container to solidify, then throw out the agar in solid waste.
7. Leave the pour plates on the bench until solidified. The plates will harden faster if they are not stacked but are left sitting on the lab bench. The color of the media will be lighter and more opaque when the medium solidifies.
8. After the agar solidifies, invert the plates and incubate at 35°C for 48 h.

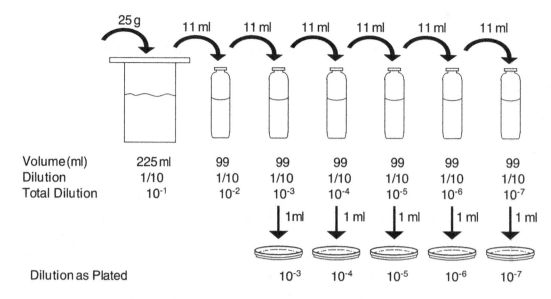

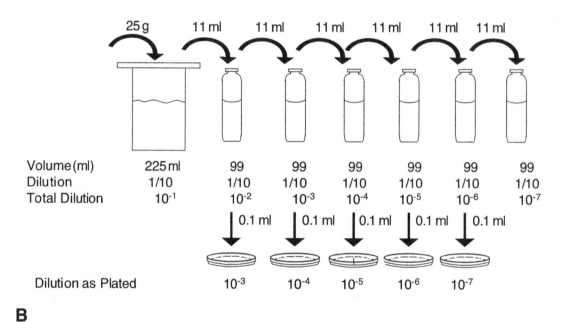

Figure 1.3 Dilution schemes for Laboratory 1: A. Dilution scheme for pour plates. B. Dilution scheme for spread plates. All dilutions should be plated in duplicate.

Spread Plate Method

Procedure

1. Label each plate count agar (PCA) petri plate with names, spreads, and dilutions as plated. We will be plating each dilution in duplicate according to the dilution scheme in Figure 1.3B.

2. Pipette 0.1 ml of the appropriate dilution onto each plate. Remember that this will be an additional 1/10 dilution.

3. The glass spreaders are affectionately called "hockey sticks." Ethanol (70%) will be used to sterilize the hockey stick. Before you start, clear the benchtop around the burner and ethanol of flammable materials to assure that burning ethanol does not drop onto your lab notebook or other objects. Place the glass spreader into 70% ethanol. Take the stick out of the ethanol, and allow the excess ethanol to drip off. Holding the handle higher than the spreading surface, quickly bring the glass through a Bunsen burner flame.

NOTE: **Danger! The ethanol will ignite. Never place a burning spreader back into your beaker of ethanol. If you accidentally ignite your ethanol, do not touch it. Notify your instructor immediately.**

4. After all the ethanol has burned off the hockey stick, it is ready to use.

5. Use a preflamed or sterile hockey stick to spread the 0.1 ml of sample evenly around the agar surface and to all edges. Keep spreading until the liquid is no longer visible on the surface.

   Theoretically, one sterile hockey stick can be used to spread a whole series of dilutions, as long as you start from the most dilute to the least dilute. However, if any plates are contaminated with surface growth, you may inadvertently contaminate all higher dilutions. Personally, I like to sterilize the hockey stick once for each dilution and use it to spread duplicate plates.

6. Invert the plates, and incubate at 35°C for 48 h.

Class 2

Results from Pour Plates

1. Check the agar control plate to make sure the agar was not contaminated.

2. Look carefully at the plates. Colonies may be on the surface, but the majority will be embedded in the agar. Embedded colonies will usually be shaped like footballs or stars, and they can be small. Be careful you do not count food particles.

3. Use an illuminated plate counter with a magnifying glass to help you see the colonies. Place the petri dish lid-side down, and use a permanent marker to mark each colony as you count. If the plates are crowded, a handheld tally can help you keep track of the numbers of counted colonies.

4. It is often difficult to see the colonies where the agar meets the plate side. Holding the plate up to a light at an angle can help you see these colonies.

5. Follow the guidelines described above for calculating the count for your fish sample. Record your results in the Laboratory 1 results page.

Results from Spread Plates

1. Look carefully at the plates. Colonies should be growing on the surface. All colonies are counted, regardless of size. (Even pinpoint-sized colonies should be counted.)

2. As described above, invert the plate and use a permanent marker to mark each colony as you count. If needed, an illuminated plate counter can be used.

3. Count plates according to the counting rules, and use these data for calculating CFU/g. Record your results in the Laboratory 1 results page.

Mechanical Dilution — Streak Plate

Note: **Streak plating takes practice. In class, each person should streak two plates using the parallel line quadrant streak and the undulating line quadrant streak.**

The objective of streaking a plate is to obtain isolated colonies. There are many types of streak patterns. The two described here are both quadrant streaks with slightly different patterns. Try the different streak patterns described below, and decide which pattern is your personal preference.

First Strip of a Quadrant Streak Plate
This is the same for both streaking techniques (Figure 1.4).

Procedure

1. Place a petri dish containing agar inverted on the bench. Label the bottom with date, researcher, and any pertinent information.
2. Pick up an inoculating loop or needle.
3. Place loop into the blue flame of the Bunsen burner until the wire is red hot (sterilization via heat). Heat the end nearest the handle first and the end with the loop last. Heating in this manner will help prevent splattering if the loop contains culture.
4. If streaking from broth, one hand should hold the test tube, and the second hand should hold the loop and the tube cap. Open the tube of broth (take cap off using the pinkie finger and palm of one hand), flame the opening of the tube (quickly pass through the flame), and place the loop into the broth.
   a. If streaking from a spread plate colony, take the hot loop and cool it on an unstreaked portion of the plate. Touch the loop to an isolated colony.
   b. If streaking from a pour plate colony, take a hot needle and cool it on an uninoculated portion of the plate. Stab through the agar and touch the isolated colony.
5. Firmly pick up the agar side of the petri dish to be streaked by cradling it with your fingers over the palm of your other hand. Adjust your wrist so you can see light reflected from the surface of the agar. Holding it in this manner will also help prevent gouging of the agar surface.
6. Place loaded loop (or needle) on the open agar plate, and start at the first pass (1) of streak (Figure 1.4A and Figure 1.4B). Gently drag the loop back and forth over the same region of the petri dish. This first pass will be the heavy growth (or lawn), which you will dilute in subsequent streaks. This step is the same regardless of the streak pattern.

Parallel Line Quadrant Streak Technique
This method uses parallel lines to physically dilute the cells. The defined pattern is sometimes easier for new students than the undulating line quadrant streak described below, although some find it more time consuming.

Procedure
See Figure 1.4A.[5]

1. Flame the loop to red hot again. Place the loop in a nonstreaked portion of the agar to cool.
2. Streak four to five parallel streaks through the inoculation area 1. Try to keep the streaks close to the side of the plate. This is now area 2.
3. Flame the loop to red hot, and place the loop in a nonstreaked portion of the agar.
4. Streak six to seven parallel lines from area 2. Once again, try to keep these lines close to the side of the plate. This is now quadrant 3.

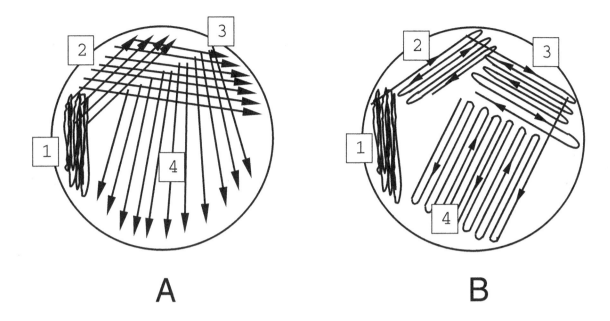

Figure 1.4 Steps involved in streaking a plate for isolated colonies: A. Parallel line quadrant streak. B. Undulating line quadrant streak. After each streak step (1 through 4), the inoculating loop or needle should be flamed to red hot and allowed to cool before proceeding to the next step.

5. Flame the loop again, and cool in a nonstreaked portion of the agar. Streak as many lines as you can from area 3. Try to make one streak go through all of the streaks from area 3, one go through six out of seven, one go through five out of seven, etc. Fill as much of the plate as possible.

6. Invert the plate, and incubate in an air incubator.

Undulating Line Quadrant Streak

Procedure
See Figure 1.4B.

1. The key to this method is to move the loop back and forth as many times as possible during the streaking of quadrants 3 and 4. The premise is to maximize the length of each streak by moving back and forth along the agar surface. The movement of the loop should be similar, using a pencil to lightly shade an area when drawing on paper.

2. Flame the loop to red hot, and allow it to cool.

3. Pull cells from the inoculation area, and streak a wavy line away from the inoculation area 1. Try to go into the inoculation area only once or twice. This is now area 2.

4. Flame the loop to red hot, and allow it to cool.

5. Go into area 2 once and pull cells along the side of the plate. Proceed with streaking a wavy line to form quadrant 3.

6. Pull cells from area 3 and make a wavy streak to fill the remainder of the plate.

7. Invert the plate, and incubate in an air incubator.

Select two colonies from your spread plate or surface colonies from your pour plates to purify by streak plating. After trying each method once, try to separate two different colony types. Pick two colonies that are obviously different (for example, one pigmented and one nonpigmented) on the same sterilized loop. Perform your streak. Determine if you were able to isolate colonies from each.

Class 3

Observe streak plates. Do all the colonies look the same? Things to look for are differences in colony morphologies (size, color, transparency, and shape).

■■■ IV. RESULTS

Spread Plate

Dilution as Plated	CFU/Plate	CFU/Plate	Average CFU/Plate
10^{-3}			
10^{-4}			
10^{-5}			
10^{-7}			

CFU/g =

Pour Plate

Dilution as Plated	CFU/Plate	CFU/Plate	Average CFU/Plate
10^{-3}			
10^{-4}			
10^{-5}			
10^{-7}			

CFU/g =

Describe the morphologies of the colonies selected for streaking.

Colony A

Colony B

Colony C

Colony D

Which streaking method was used? Was it successful in isolating isolated colonies?

Colony A

Colony B

Colony C

Colony D

■ V. DISCUSSION QUESTIONS

1. Were the numbers you observed with the spread plate vs. pour plate different?
2. What errors are associated with SPC methods? Describe some things that could contribute to problems with reproducibility.
3. For SPCs, PCA or tryptic soy agar (TSA) is used. Why do you think the protocol for SPC from fish requires the addition of 0.5% NaCl to the agar? How would you expect the results to change if 5% NaCl were added? How would you expect results to change without additional NaCl?
4. Why are the counting rules important to follow? Describe why it is important to remember that each colony is a CFU rather than a single organism?
5. Did your colonies on the streak plate appear homogeneous after streaking? Using techniques from this lab, how could you further purify these cultures, and how would you confirm their purity?
6. One semester we had two sections of this lab back to back in the morning. The first section was two hours before the second. The SPC results from the first section (for both pour and spread) ranged from 1×10^4 to 5×10^5 CFU/g and were noticeably lower than the results from the second section (which ranged from 3×10^6 to 7×10^7 CFU/g). What are some potential explanations for why the second class had consistently higher cell numbers? Assuming it is not laboratory technique, why would CFU/g levels on the same fish fillet vary by over 1 log unit?

■ LABORATORY NOTES

■ LABORATORY NOTES

■ LABORATORY NOTES

LABORATORY 2

MICROSCOPIC EXAMINATION
OF YEAST, MOLD, AND BACTERIA

■ I. OBJECTIVES

- Learn the difference between simple and differential stains.

- Become familiar with staining and observing yeast.

- Examine different molds, and identify morphologies of mycelium and hyphae.

■ II. BACKGROUND

Bacterial Mount

Prior to observing anything under a microscope, the sample must be prepared on a microscope slide. This is known as mounting. For a standard compound light microscope, bacteria are usually prepared as a dried film, which is heat fixed and stained. Samples do not have to be stained for observation using a phase contrast microscope; therefore, a simple wet mount can be used.

Simple Stain

A simple stain will stain all cells in a sample the same color, which aids in observing cellular morphology or performing a direct microscopic count of cell numbers. Methylene blue will be used in this lab exercise. This is a basic stain. Basic stains tend to have a high affinity toward acidic cell wall components.[6]

Gram Stain

This stain is named for Christian Gram, who developed it in the 1800s. This stain differentiates between two broad classes of organisms, which differ in their cell wall compositions. Gram-positive (Gram+) cells have thick peptidoglycan layers. Gram-negative (Gram-) cells have thin peptidoglycan layers and a lipopolysaccharide (LPS) layer.

In this staining procedure, crystal violet is the primary stain. Like methylene blue, this stain is alkaline and has a high affinity to cell components. The mordant is iodine. The purpose of the mordant is to combine with the dye to form an insoluble complex between the dye and cellular components.[6] This complex, when present on certain cells (Gram+), will be resistant to the decolorization step. In the decolorization step, ethanol is used to remove excess dye from the slide as well as dye from certain cells (Gram-). The counterstain, safranin, will be used to stain decolorized cells a reddish pink color.

The Gram stain is a step-by-step procedure. Reproducibility is based upon strict adherence to staining times and careful attention to technique. Errors can be incorporated by poor techniques, as explained in the following examples[6]:

1. Overheating the film while heat fixing: If the cells break open due to overheating, the Gram+ cells may lose the crystal violet complex and appear as Gram-.

2. Too many cells in the film: This may cause irregular staining and decolorization may be incomplete. This could cause Gram- cells to appear as Gram+.

3. Decolorization is the critical step: If too much time is taken, you will decolorize the Gram+ cells. If too little time is taken, you will not completely remove the crystal violet from the Gram- cells.

4. Age of the culture: Old cultures may not stain accurately.

5. Iodine solution can deteriorate over time: The solution should be dark golden yellow in color.

Microscopic Observation of Fungi

Filamentous fungi are usually identified by basic structures that can be observed by light microscopy. Filamentous fungi have long filaments or tubes called "hyphae." Within the hyphae, fungi may or may not have "septa" or cross-walls. Intertwined hyphae are called a "mycelium." Within the mycelium, a portion remains on the substrate, and reproductive shoots grow into the air. Filamentous fungi can grow asexually (imperfect form) or sexually (perfect form). Within a fungi genus, the structures will vary between the perfect and imperfect forms. Figure 2.1 represents the nonsexual growth of fungi that you will be looking at in this lab.

The genera *Aspergillus* and *Penicillium* belong to the Deuteromyces phylum of fungi. All members of this group have branching septate hyphae and rarely have a perfect (sexual) form. They both produce "conidiospores" on top of aerial structures known as "conidiophores." The conidiospores are produced in long chains and are attached to the conidiophore via a structure called a "phialid" (Figure 2.1). In general, members of the genus *Penicillium* have a narrow head of conidiospores, while *Aspergillus* members usually have a spherical head of conidiospores. The "foot cell" is unique to the genus *Aspergillus*, and this three-pronged structure is found at the base of the conidiophore where it meets the mycelium (Figure 2.1).

Mucor and *Rhizopus* are genera of the Zygomycetes class of fungi. These organisms are nonseptate and produce asexual "sporangiospores" or sexual "zygospores." The fruiting bodies of these two genera are similar: the aerial structure is called a "sporangiophore" with a large globular "sporangium" that contains large amounts of sporangiospores under a membrane. When mature, the spores are released after the rupture of the sporangial membrane. *Mucor* and *Rhizopus* are differentiated based upon the organization of the mycelium. Generally, members of the *Mucor* genus develop sporangiophores randomly in any direction from the branching mycelium (Figure 2.1). Members of the *Rhizopus* genus have a more organized structure, with nodes in which the aerial sporangiophores grow up and structures called "rhizoids" develop growing down (Figure 2.1). Some members of *Rhizopus* develop with the rhizoids positioned directly below the sporangiophores (as pictured in Figure 2.1), and other members may have the rhizoids growing downward elsewhere from the mycelium.

To get the best view of fungi structures, they should be propagated on a small amount of solid media mounted on a glass slide. In this lab, we will be using a quick method, or "the old transparent tape trick," to obtain enough hyphae with which to observe the fungal structures.

Use of Light Microscopes

The majority of microscopes used in teaching laboratories have three to four objective lenses and an eyepiece. The eyepiece magnifies by 10×, and each objective is marked with its degree of magnification (typically, 10×, 40×, and 100×). To determine the total magnification, multiply your objective magnification by the eyepiece magnification. The 100× objective is typically an oil immersion lens (although some scopes have 60× oil immersion lenses). The oil has a lower refractive index than air, thus sharpening the image seen at the higher magnifications.[7]

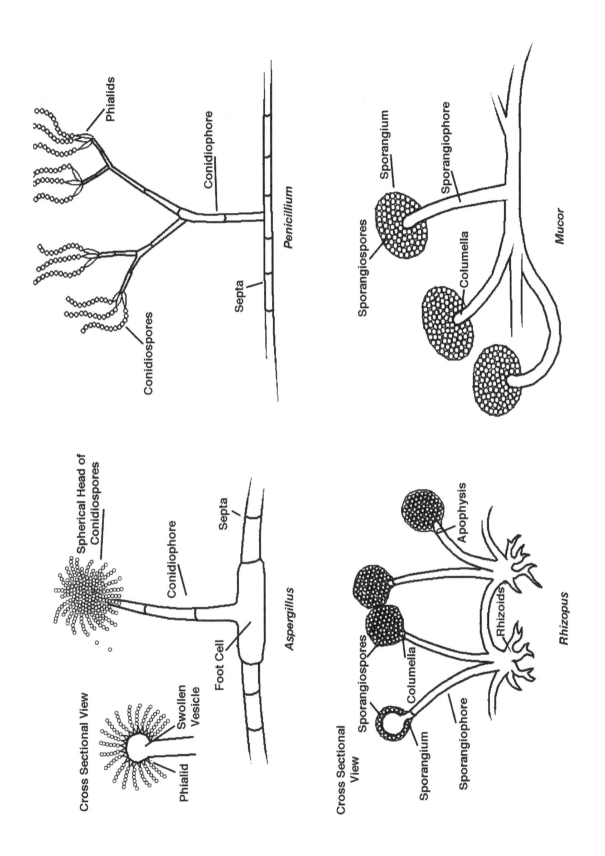

Figure 2.1 Diagram of fungi used in this lab.

General Microscope Use

(The following is a general description. Make sure you check with your instructor prior to using your microscope.)

Note: **Remember to always move the objective and stage away from each other when looking through the eyepiece. If you need to bring them closer together, always take your eye away from the eyepiece and watch the movement from the side. This will prevent lens damage.**

1. Rotate the nosepiece until the lowest power (10×) objective is in the viewing position. The lower the power of the objective lens, the greater the area of specimen surface is included in the field of view with a greater depth of focus. For these reasons, the lower objectives are always used for initial focusing and viewing before changing to oil immersion (100×).

2. Plug in the power cord, and turn on the light source.

3. Place the slide on the stage between the stage "fingers" on the mechanical stage. Use the control knobs of the mechanical stage to position the specimen area of the slide over the center of the stage aperture.

4. Look directly at the slide (not through the eyepiece). Raise the stage until it reaches an upward stop. Make sure to watch from the side to make sure this stop occurs. Do not allow the objective to hit the slide.

5. Look through the eyepiece and lower the stage with the coarse adjustment (outer knob) until an image appears. When you look at bacteria, this image may only be a slight amount of blurry stain color.

6. Adjust the fine focus knob to sharpen the image and bring it into focus. Look at the image and adjust the condenser aperture (located below the stage) to obtain the clearest possible image. The clearness of the image depends upon the size of the aperture. As the aperture becomes smaller, the contrast and depth of focus increase but the resolving power decreases. The clearest image is produced by the best combination of these factors.

7. After focusing at the lowest objective (10×), you can rotate the nosepiece to a higher objective. Theoretically, the objectives should be properly aligned so that the focus does not change between objectives. In reality, you may need to use the fine focus to sharpen the image after moving to each higher objective.

8. The 100× objective is an oil immersion lens. This is the only lens that uses oil. Be careful, because using oil with any other lenses will result in major lens damage. To use the oil objective lens:

   a. After performing initial focusing (10×, 40×), gently rotate the lenses back to the 10× objective.

   b. Add a small drop of immersion oil to the lighted area on the specimen slide. Try to avoid air bubbles.

   c. Rotate the nosepiece until the 100× (oil immersion) objective is in the light path and is touching the oil.

   d. Look through the eyepiece and adjust the fine adjustment. The aperture may need to be adjusted to allow more light to come through the slide. If nothing is visible using the fine adjustment, perform the following:

   e. While watching the stage (not through the eyepiece), use the coarse focus knob to lower the objective, while watching the space between the objective and the slide. Slow down when you see the lens make contact with the oil drop (you may see a flash of light), and bring the objective and stage as close together as you can without making direct contact.

f. Looking through the eyepiece, slowly lower the stage. After focusing on your specimen, you may need to readjust the condenser aperture to achieve the greatest amount of contrast and resolution.

g. Each time you finish using the oil immersion objective, wipe off all traces of oil from the objective with special lens paper. (Do not use laboratory wipes.)

Cultures Used in this Lab

Bacterial: Broth mixed culture; use isolated colonies from streak plates (from Laboratory 1 or those provided by the instructor)

Yeast: *Saccharomyces cerevisiae*

Molds: *Penicillium* spp.; *Mucor* spp.; *Rhizopus* spp.; *Aspergillus* spp.

▄▄ III. METHODS

Preparation of "Heat-Fixed" Bacterial and Yeast Smears

You will need to prepare one heat-fixed smear of yeast and two heat-fixed smears of each bacterial culture:

1. Make a smear of a culture on a microscope slide.
   a. If from a plate, place one loop of water on the slide. Flame a loop, and select a single colony for examination. Place the colony in the water on the slide, and spread it into a thin film (approximately 1 cm²).
   b. If from broth, place one loop of culture on the slide, and spread it into a thin film (approximately 1 cm²).
2. Allow the smear to dry completely. If this is not allowed to dry, you will later lose your bacterial smear. Slides placed at the base of a lit burner will dry faster.
3. Place the slide into a clothespin. Quickly bring the slide through the flame once or twice. The slide should not be burning hot. If you overheat the slide at this step, you can change the staining characteristics of the organisms.

Methylene Blue Staining

Perform methylene blue staining on yeast and each bacterial culture.

Procedure

1. Prepare a heat-fixed film as described in the previous section.
2. Cover the film with methylene blue solution for 1 to 2 min.
3. Tilt the slide to allow the excess stain to run off into the staining tub, and wash gently with the water bottle.
4. Allow the slide to air dry without blotting or blot gently with a lint-free laboratory wipe.
5. Examine the stained yeast and bacteria with your microscope using the 40× and oil immersion objective, respectively.

Gram Stain

Gram stain the mixed culture and a colony from your streak plates from Laboratory 1 or from those provided by your instructor.

Procedure

1. Cover the heat-fixed film with crystal violet solution and leave for 1 min.
2. Tilt the slide to allow excess crystal violet to run off into a staining pan. Rinse with water bottle gently for 1 to 3 sec.
3. Make sure the excess water is drained off, then cover the film with iodine solution and leave it on for 1 min.
4. Drain the iodine solution and rinse with water as above. Allow the excess water to run off.
5. Flood the film with alcohol (95% ethanol) for less than 30 sec to remove the purple/blue color. (This step is critical and should be performed with extreme care. It can also be done by holding the slide at an angle and adding about 5 to 10 drops of the alcohol, drop-wise, until a purple/blue color no longer streams from the film.)
6. Rinse immediately with water.
7. Cover the film for 30 to 60 sec with safranin solution.
8. Rinse it with water, blot gently, and allow the film to air dry.
9. Observe the stained film under oil immersion. You may need to adjust the condenser to get the right amount of light.
10. Record the Gram reaction of each stained sample. For identification purposes, the cellular morphology (such as long, short, or irregular rods or chains, clumps or individual cocci) should be recorded for each sample.

Observation of Molds

1. Record colony characteristics, such as color and texture.
2. Use a dissecting microscope to observe the mass of mycelia at the edge of a colony.
3. Take a small piece of tape and roll it into a loose circle with the sticky edge facing out.
4. In a biological safety hood, pick up a small amount of mycelia from the edge of a colony. Uncurl the tape and place it on a microscope slide, with the mycelia (sticky side) facing down. The tape will act as a coverslip.
5. At the microscope, start at 100× (10× objective) or 20× (2× objective) magnification, and look for sections of mycelium. Tape slides will have large amounts of loose spores, so you will need to search the slide for an area with mycelium. Once you find some mycelia, look for structures described in Figure 2.1. Usually you can get a better view of filamentous fungi at lower magnifications, but you can increase the magnification to 400× (40× objective) to obtain more detail.
6. Make drawings in your notebook and label parts of each fungi.

■ IV. RESULTS

1. Record the cellular morphology of *S. cerevisiae* at 40× and 100× magnification. Look for buds and bud scars on the yeast sample.
2. Record the cellular morphology and the Gram reaction of each bacterial culture.
3. Record the colony morphology and specific morphology of hyphae and fruiting bodies of the filamentous fungi.

■ V. DISCUSSION QUESTIONS

1. What is the maximum total magnification you can obtain with your scope?

2. Focus a slide on an object (any slide, any objective). Use the mechanical stage to move the slide slightly to the left. Repeat this while looking through the microscope. What happens? Why?

3. What are the fundamental differences between Gram+ and Gram- bacteria?

4. Were you able to observe the different structures of the molds?

5. Did you see spores or buds in the yeast?

■ LABORATORY NOTES

◼ LABORATORY NOTES

■ LABORATORY NOTES

LABORATORY 3

ENUMERATION OF YEASTS AND MOLDS FROM FOODS

■ I. OBJECTIVE

- Introduce students to enumeration methods used for yeasts and molds.

■ II. BACKGROUND

There is a large diverse population of yeasts and molds that can grow on foods. They can be found on crops (grains, nuts, beans, and fruits) prior to harvest and during storage. In addition, they can be found in processed food products. In general, yeasts and molds are considered to be spoilage organisms. Some yeasts and molds, however, are a public health concern due to their production of mycotoxins, which are not destroyed during food processing or cooking.[8]

In general, most yeasts and molds require oxygen for growth. Their rates of growth are generally slower than those of bacteria; however, their growth ranges are much wider, encompassing more severe environmental conditions. Various yeasts and molds can grow over a wide pH range (around pH 2 up to pH 9) and a broad temperature range (5 to 35°C).[9] In addition, some genera can grow at reduced water activities ($a_w \leq 0.85$).[9]

Selective media and lower incubation temperatures are used to slow or inhibit bacterial growth and thereby selecting for growth of yeasts and molds. Selective bacterial inhibition can be achieved using antibiotics (such as chloramphenicol at 100 µg/ml or gentamicin 50 µg/ml) or through acidification of media (acidification of potato dextrose agar with tartaric acid to pH 3.5 is often used).[8,9] To inhibit colony spreading and excessive mycelia formation, dichloran (2 µg/ml) and/or rose bengal (25 µg/ml) are a common additive to mycological count media. For yeasts and molds recovered from foods with reduced water activities ($a_w \leq 0.85$), media containing 18% glycerol (final $a_w = 0.995$) are recommended to assist in recovering organisms that may be sensitive to environments with high water activities.[8]

In this laboratory, you will perform a yeast and mold count using fresh refrigerated salsa.

■ III. METHODS

The methods discussed here were adapted from the literature.[9]

Yeast and Mold Count

Procedure

1. Measure 25 g of refrigerated salsa and add to 225 ml 0.1% peptone water in a sterile Stomacher bag. Homogenize for 1 min in a Stomacher blender.
2. Follow the dilution scheme outlined in Figure 3.1.
3. Spread plate in duplicate on plate count agar with 100 μg/ml chloramphenicol or potato dextrose agar acidified to pH 3.5 with tartaric acid. Incubate in an upright position at 22 to 25°C for 5 days.
4. Count the plates containing 15 to 150 colonies. The lower numbers are used because colonies are larger than bacteria. Record the results on the results page. Because mold allergies are common, do not open petri plates unless in a biological safety hood.

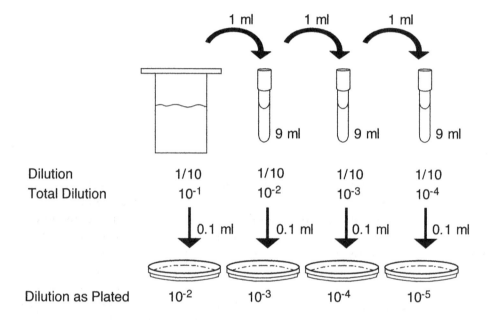

Figure 3.1 Dilution scheme for spread plating of diluted salsa. All dilutions should be plated in duplicate.

■ IV. RESULTS

Dilution as Plated	CFU/Plate	CFU/Plate	Average CFU/Plate
10^{-2}			
10^{-3}			
10^{-4}			
10^{-5}			

CFU/g =

■ V. DISCUSSION QUESTIONS

1. How is this count selective for yeasts and molds?
2. Were the colonies on your plates distinct and separate or were they spreading? What can be added to the media to reduce the amount of spreading colonies?
3. What added media are used to recover molds from foods with reduced a_w? Why is this done?

■ LABORATORY NOTES

■■ LABORATORY NOTES

■ LABORATORY NOTES

_____

_____

_____

_____

_____

_____

_____

_____

_____

_____

_____

_____

_____

_____

_____

_____

_____

_____

_____

LABORATORY 4

COLIFORMS AND *ESCHERICHIA COLI* FROM WATER: MOST PROBABLE NUMBER METHODS AND 3M® PETRIFILM™

■ I. OBJECTIVES

- Become familiar with enumeration by the traditional most probable number (MPN) technique.

- Become familiar with differential and selective bacteriological media.

- Become familiar with 3M Petrifilm (St. Paul, MN).

■ II. BACKGROUND

Differential and selective media are used extensively to differentiate between different groups, genera, and species of bacteria and will be used in almost all subsequent labs:

- A differential media contains biochemical substrates that may or may not be utilized (or modified) by different bacteria. The utilization or modification of the substrate is usually indicated by a color change in the colony and/or surrounding bacteriological media. Examples of commonly used differential reactions are sugar utilization (with a pH indicator and possibly an inverted tube to trap gas) or enzymatic reactions (using a fluorescent substrate or clearing an added agent from solid media).

- A selective media contains one or more chemicals that reduce or inhibit the growth of interfering or "background" organisms and allow the visualization of the target organism. Selective agents include inorganic salts, dyes, surface active agents (such as bile salts), and antibiotics. However, it is important to note that target organisms normally resistant to selective agents may become more sensitive (and growth inhibited) if cells were injured or stressed.

Note: **Most media use a combination of selective and differential agents to reduce the growth of background organisms and allow the visualization of the target organism.**

In both food and water analyses, it is too expensive and time consuming to analyze every sample for every pathogen, and therefore, microbial analysis often utilizes "indicator organisms." Indicator organisms are usually present in higher numbers than pathogens and are easier to detect. In addition, indicator organisms should have growth and survival characteristics that are similar to those of pathogens. The most commonly used indicator organisms are coliforms, which comprise a loosely defined group of Gram-negative (Gram-) organisms that are members of the *Enterobacteriaceae* family, most of which can be found in the intestines of warm-blooded animals. Because many foodborne pathogens are also members of the *Enterobacteriaceae* family and shed through the intestinal tract, the presence of high numbers of coliforms can be used to predict intestinal pathogens. High levels of coliforms can also indicate the absence of sanitation.

Genera included in the "coliform" term include at least four: *Escherichia, Klebsiella, Citrobacter,* and *Enterobacter.* The definition of coliforms is a laboratory definition based upon Gram stain and metabolic reactions. By definition, coliforms are: "Gram-negative, non-spore forming, aerobic or facultative anaerobic rods that ferment lactose, forming acid and gas within 48 hours at 35°C."[1]

E. coli is a member of the "coliform" group. When enumerating coliforms, *E. coli* is among the mixed population measured. However, some selective or differential media can be used for evaluating *E. coli* levels independent from the mixed coliform population. Counts of *E. coli* are more specific than coliform counts in that there is only one species detected. In addition, because a small population of *E. coli* serotypes can be pathogenic, some researchers believe that the presence of *E. coli* may have a more accurate correlation to the presence of pathogens than coliforms.

The MPN technique is a statistical method that is useful in determining low concentrations of organisms (<100/g or ml).[4] In this method, samples are serially diluted so that the inocula will sometimes, but not always, contain viable organisms. At each dilution, multiple volumes are transferred into three, five, or ten tubes of liquid test medium. The tubes are incubated, and results are evaluated. Depending on the test performed, positive tubes may be identified by turbidity (growth) alone, or in combination with gas and acid production. In some cases, biochemical substrates may also be used. After grading tubes positive (+) or negative (-), the initial MPN/g (or ml) levels are determined with a MPN table. This table will give you the estimated statistical level and the 95% confidence intervals associated with this number. As with anything involving statistics, as you increase the number of tubes (samples) at each dilution, the accuracy of the method will improve. So, a ten-tube MPN will be more accurate than a five-tube MPN, etc. At a certain point, the amount of media needed for multiple tube analysis becomes cost prohibitive.

The MPN technique is based upon the following assumptions:

- Bacteria are distributed randomly within the sample.

- Bacteria are not clumped or clustered together.

- The growth medium and incubation conditions will allow for detectable growth from even one organism.

Petrifilm was introduced about 20 years ago and is currently available in a number of different formulations. Petrifilm is essentially miniaturized dehydrated petri plates that contain a cold gelling agent. The sample (1 ml) is plated, and the liquid hydrates the media. The major advantages to Petrifilm are as follows:

- There is no need for media preparation and sterilization.

- It is compact in size. (Twenty films take up incubator space approximately equivalent to two petri dishes).

- Gas bubbles can be detected after incubation. (This is especially useful in identifying coliforms using the laboratory definition).

Disadvantages include the following:

- The small colony size makes counting more difficult.

- Food particles may imitate colonies when small dilutions are plated.

For traditional MPN identification of coliforms and *E. coli*, we will use lauryl tryptose broth supplemented with a fluorescent substrate 4-methylumbelliferyl-β-D-glucuronide (MUG). About 90% of all *E. coli* produce the enzyme β-D-glucuronidase, although other organisms (such as some strains of *Erwinia*) can also produce this enzyme. This enzyme will cleave MUG and produce a fluorescent product that can be visualized with long-wave ultraviolet (UV) light, allowing for the identification of presumptive tubes containing *E. coli* growth. The presence of β-D-glucuronidase activity is evident when using *E. coli* Petrifilm, but instead of MUG, a colorimetric substrate is used, and a blue precipitate surrounding each colony is an indication of presumptive *E. coli* β-D-glucuronidase activity.

Use of the MPN Table

The MPN table presented is from the U.S. Food and Drug Administration's *Bacteriological Analytical Manual* (FDA BAM, 1999).[4] This table is set up for tubes with 0.1, 0.01, and 0.001 g inoculum levels. We will use 1 ml of 10^0, 10^{-1}, 10^{-2}, and 10^{-3} dilutions. This translates into 1, 0.1, 0.01, and 0.001 ml volumes of inoculum in the series:

1. Select three dilutions for table reference. This is the trickiest part of performing an MPN.
   a. Select the most dilute tubes (highest dilution) with all positive replicate tubes and the next two higher (more dilute) dilutions.
      i. For example, for our dilution scheme, if you had the number of positives as 10^0 3/3, 10^{-1} 3/3, 10^{-2} 1/3, and 10^{-3} 1/3, you would use 3-1-1 in the MPN chart.
   b. If there are not two higher (more dilute) dilutions available, then select the three highest dilutions.
      i. For example, if you had 10^0 3/3, 10^{-1} 3/3, 10^{-2} 3/3, and 10^{-3} 1/3 as your results, then you should use 3-3-1 for the MPN chart.
   c. If no dilutions show all positive tubes, select the three lowest (least dilute) with a positive result.
      i. For example, if our results were 10^0 0/3, 10^{-1} 1/3, 10^{-2} 0/3, and 10^{-3} 0/3, you would use 0-1-0 for the MPN chart. However, the MPN number would have to be divided by 10, because you used a 1, 0.1, 0.01 series with a 0.1, 0.01, 0.001 MPN chart. This will be true whenever you use the 10^0 results.
2. After you select your three dilutions, use the chart (Table 4.1) to determine the MPN/ml.
3. If using the 10^{-1} to 10^{-3} dilutions, use this number directly from the chart. If you are using the 10^0 to 10^{-2} dilutions, you must divide the MPN by 10, because this is a 0.1, 0.01, 0.001 g MPN chart.

Bacteriological Media Used in this Lab

1. Lauryl sulfate tryptose (LST) broth with MUG: This medium contains tryptose (nitrogen source), lactose (carbohydrate), phosphate buffer, salt (NaCl), lauryl sulfate, and MUG.

   Selective agent: Sodium lauryl sulfate, which inhibits growth of Gram-positive (Gram+) and spore-forming bacteria.

   Differential agents: (a) Gas production from lactose (duram tube) and (b) fluorescent by-product due to cleavage of MUG by *E. coli*, which produce the enzyme β-D-glucuronidase.

2. Brilliant green bile (BGB) broth: This medium contains peptone (nitrogen source), lactose (carbohydrate), oxgall, and brilliant green.

   Selective agents: Brilliant green and oxgall — both inhibit the growth of Gram+ organisms.

   Differential agents: (a) Gas production (duram tube) and (b) acid production — brilliant green turns yellow at around pH 4.[6]

3. *E. coli*/coliform Petrifilm: This medium is a proprietary modification of violet red bile agar (VRBA)* and most likely contains yeast extract, peptone, bile salts, lactose, sodium chloride, neutral red, and crystal violet.

>   Selective agents: Bile salts and crystal violet inhibit the growth of Gram+ organisms.

>   Differential agents: (a) Lactose-positive organisms have purplish red colonies. (b) Lactose-negative organisms are clear to pink. The substrate for β-D-glucuronidase changes from colorless to a blue precipitate when cleaved.

▬ III. METHODS

MPN with LST Broth + MUG

In this lab, students will perform an MPN enumeration of coliforms with LST broth + MUG. LST broth contains lactose as the carbohydrate source. Tubes positive for coliforms produce gas while metabolizing lactose in LST broth. Growth from these tubes must then be transferred into BGB broth to confirm acid production from lactose. We will be able to estimate presumptive coliform MPN/ml and presumptive *E. coli* MPN/ml. Growth from tubes that are presumptive positive for coliforms (turbidity and gas) will be transferred to BGB broth. Production of acid and gas will confirm the coliform-positive tubes, and these numbers can be used to calculate the confirmed coliform MPN/ml. The LST broth will also be used to give us a presumptive *E. coli* count. *E. coli*-positive tubes will have turbidity and gas and will fluoresce under UV light.[4]

Class 1

MPN Analysis

Procedure

1. Dilute an environmental water sample as shown in Figure 4.1.
2. Transfer 1 ml volumes into a series of three tubes. The three tubes do not need to be differentiated, just mark them with the dilutions.
3. Incubate tubes at 35°C. Read the tubes after 48 h.

Plating upon E. coli Petrifilm

Procedure

1. Use the dilution series that you prepared for the MPN.
2. Label each petri film with the dilution as plated. Because 1 ml of each serial dilution will be used, the final dilution would be the same as the tube dilution (for example, 1 ml × 10⁻² is equal to a final dilution 10⁻²).
3. Place Petrifilm on the bench, and plate one dilution at a time.
4. Pull up the top film, and add 1 ml of the dilution to the dehydrated media.
5. Roll down the film to prevent air bubbles.
6. Use plastic template (ridge side down) to gently press down the sample to fill the area. (Be careful not to press so hard that the sample sloshes outside the area of media.)
7. Incubate plates 35°C for 24 h (as suggested by the manufacturer).

---

* VRBA can be purchased and used as a traditional agar. Traditional VRBA is somewhat complicated to make and requires an overlay of a small amount of tempered agar after spread plating. Lactose-positive acid colonies (deep red) then must be transferred into liquid media (LST or other lactose broth) to confirm that gas is produced during lactose fermentation. A major advantage of using Petrifilm is that the lactose fermentation and gas production can be done at the same time to reduce the amount of time needed for enumeration of coliforms.

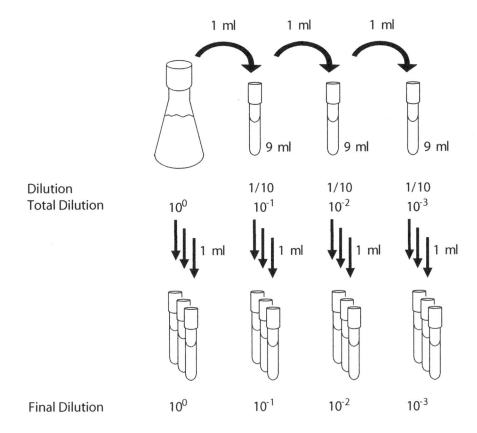

Figure 4.1 Dilution scheme for MPN from an environmental water sample. The same dilution should be used for the Petrifilm plating in duplicate.

Class 2

Results for MPN

1. Record the tubes with turbidity and gas production as presumptive coliform positive in the results section. Use MPN tables to determine the presumptive MPN of coliform/ml after 48 h.

2. Transfer a loopful of growth from all tubes showing gas production into a tube of BGB broth for confirmation of coliforms. Incubate for 48 h at 35°C.

3. Expose the tubes to UV light. (**Caution:** Eye protection needs to be worn during UV light exposure.) Record the tubes that fluoresce as *E. coli* positive.

Results of E. coli Petrifilm

1. Count films with 25 to 250 CFU/ml and those that do not have excessive gas formation. Plates containing excessive gas formation are designated as too numerous to count (TNTC).

2. Coliforms will ferment lactose (red colonies) and produce gas (entrapped bubbles). Colonies associated with white foam are not counted as coliforms. Count the red and blue colonies and use the data to calculate coliform CFU/ml.

3. Count colonies that have a blue coloration and gas bubbles as *E. coli*. Use these numbers to calculate *E. coli* CFU/ml.

Class 3

Confirming MPN

1. Tubes with acid and gas production in BGB broth should be considered confirmed coliforms.

2. Use confirmed data in the MPN table to calculate confirmed MPN coliform/ml. Select the appropriate three dilutions for the chart (see Section II, "Background") and use the chart (Table 4.1) to determine the MPN/ml.

3. If using the 10^{-1} to 10^{-3} dilutions, use the MPN number directly from the chart. If using the 10^{0} to 10^{-2} dilutions, you must divide the MPN by 10, because this is a 0.1, 0.01, 0.001 g MPN chart.

◼◼ IV. RESULTS

Presumptive Coliforms MPN/ml

	Positive Tubes				MPN/ml (Table 4.1)
	10^0	10^{-1}	10^{-2}	10^{-3}	
Presumptive coliform					Na[a]
Confirmed coliform					
E. coli					

[a] Na = not applicable.

Coliform Count on Petrifilm

Dilution as Plated	CFU/Plate	CFU/Plate	Average CFU/Plate
10^0			
10^{-1}			
10^{-2}			
10^{-3}			

Coliform CFU/ml =

E. coli Count on Petrifilm

Dilution as Plated	CFU/Plate	CFU/Plate	Average CFU/Plate
10^{-1}			
10^{-2}			
10^{-3}			

E. coli CFU/ml =

▬ V. DISCUSSION QUESTIONS

1. Why was our MPN in LST broth only presumptive? Why did growth in BGB broth confirm the presence of coliforms?

2. List some advantages of using Petrifilm compared to the traditional VRBA media. Be sure to discuss the time factor.

3. Use our dilutions and the MPN table to determine the minimum concentration of coliforms that we could have detected with the dilutions we used for our traditional MPN. What was the maximum?

4. What was the minimum concentration of coliforms we could detect with Petrifilm? Use our dilutions and assume a minimal count of 25 coliforms/plate. What was the maximum concentration of coliforms we could detect with the dilutions used, assuming a maximum of 250 coliforms/plate?

5. Compare the results from the Petrifilm and the MPN. Were similar numbers of coliforms and *E. coli* detected with each method? Discuss potential reasons for differences.

TABLE 4.1

MPN Table for Three-Tube MPN with 0.1, 0.01, and 0.001 g Inocula

Positive Tubes				Confidence Limits		Positive Tubes				Confidence Limits	
0.1	0.01	0.001	MPN/g	Low	High	0.1	0.01	0.001	MPN/g	Low	High
0	0	0	<3.0	—	9.5	2	2	0	21	4.5	42
0	0	1	3.0	0.15	9.6	2	2	1	28	8.7	94
0	1	0	3.0	0.15	11	2	2	2	35	8.7	94
0	1	1	6.1	1.2	18	2	3	0	29	8.7	94
0	2	0	6.2	1.2	18	2	3	1	36	8.7	94
0	3	0	9.4	3.6	38	3	0	0	23	4.6	94
1	0	0	3.6	0.17	18	3	0	1	38	8.7	110
1	0	1	7.2	1.3	18	3	0	2	64	17	180
1	0	2	11	3.6	38	3	1	0	43	9	180
1	1	0	7.4	1.3	20	3	1	1	75	17	200
1	1	1	11	3.6	38	3	1	2	120	37	420
1	2	0	11	3.6	42	3	1	3	160	40	420
1	2	1	15	4.5	42	3	2	0	93	18	420
1	3	0	16	4.5	42	3	2	1	150	37	420
2	0	0	9.2	1.4	38	3	2	2	210	40	430
2	0	1	14	3.6	42	3	2	3	290	90	1000
2	0	2	20	4.5	42	3	3	0	240	42	1000
2	1	0	15	3.7	42	3	3	1	460	90	2000
2	1	1	20	4.5	42	3	3	2	1100	180	4100
2	1	2	27	8.7	94	3	3	3	>1100	420	—

Source: Adapted from the U.S. Food and Drug Administration, Center for Food Safety & Applied Nutrition, *Bacteriological Analytical Manual Online*, 2001 (http://www.cfsan.fda.gov/~ebam/bam-toc.html).

LABORATORY NOTES

■ LABORATORY NOTES

LABORATORY 5

GROUND BEEF MICROFLORA: SPC AND *ESCHERICHIA COLI* COUNT

■ I. OBJECTIVE

- Become familiar with the isolation of coliforms and *E. coli* from ground beef.

■ II. BACKGROUND

In this laboratory exercise, we will be looking at the microflora of ground beef. After a large outbreak of *E. coli* O157:H7 infections from tainted hamburgers, the U.S. Department of Agriculture (USDA) implemented a mandatory Hazard Analysis and Critical Control Point (HACCP) regulation for raw meat products. This regulation consists of development of a HACCP plan with end product testing for *Salmonella* and generic *E. coli*. In this lab, we are going to look at the CFU/g (standard plate count [SPC]) and perform a generic *E. coli* count using a selective and differential medium (violet red bile agar + 4-methylumbelliferyl-β-D-glucuronide) to count coliforms and *E. coli*.

Bacteriological Media Used in this Lab

Tryptic Soy Agar
Tryptic soy agar (TSA) is a nutrient-rich agar that will support the growth of many different organisms. This medium is not selective or differential and is appropriate for use in SPCs.

VRBA + MUG
This medium is known as violet red bile agar (VRBA) supplemented with 4-methylumbelliferyl-β-D-glucuronide (MUG). This medium is used to selectively plate for coliform bacteria and to screen for *E. coli*. The VRBA contains yeast extract, peptone, bile salts, sodium chloride, lactose, neutral red, and crystal violet. The bile salts and crystal violet are selective agents that inhibit Gram-positive organisms. The neutral red is a differential agent that indicates lowered pH due to acid production from lactose. Presumptive coliforms will ferment lactose and form purplish red subsurface colonies, 1 to 2 mm in diameter, and are generally surrounded by a reddish zone of precipitated bile. About 93 to 97% of all *E. coli* produce the enzyme β-D-glucuronidase. This enzyme will cleave MUG and produce a fluorescent product, which can be visualized with long-wave UV light. Presumptive *E. coli* colonies will look like all other coliforms but will fluoresce with long-wave UV light. MUG is an example of a differential agent.

III. METHODS

Class 1

Preparing Sample

Procedure

1. Use a sterile utensil to measure 25 g of the ground beef into a paddle blender bag.
2. Add 225 g peptone water. (This is your first 1/10 dilution, or 10^{-1}.)
3. Stomach for 2 min.
4. Prepare serial dilutions (Figure 5.1).

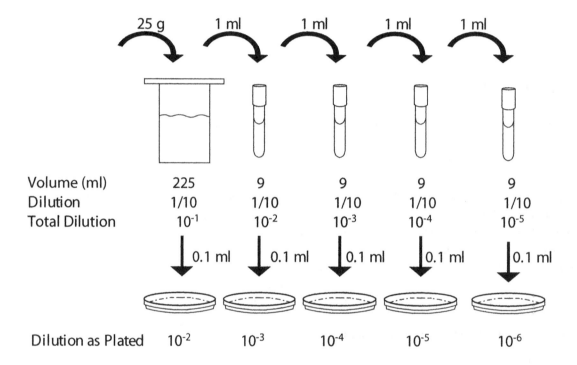

Figure 5.1 Dilution scheme for VRBA and SPC: For VRBA, spread plate dilutions 10^{-2} to 10^{-4}; for SPC, pour plate dilutions 10^{-3} to 10^{-6}.

Use pour plates for SPC and spread plates for VRBA plating. The same dilution bottles can be used to plate with plate count agar (PCA) and VRBA, but note that different dilutions will be used for each analysis. The dilution scheme is shown in Figure 5.1.

SPC Pour Plate

Procedure

1. Label an empty petri dish (group names, plated dilution). We will be plating each dilution in duplicate.
2. Pipette 0.1 ml of the 10^{-3} to 10^{-6} dilutions (Figure 5.1) into each plate.
3. Once all samples are placed in the petri dishes, get one bottle of tempered (45 to 47°C) liquid agar PCA from the water bath. Each bottle has approximately 100 ml of agar that will be enough for four pour plates (25 ml agar/plate).

4. Gently pour in tempered agar while swirling the plate gently. Remember to drip any remaining agar from both bottles into a media control plate and incubate it alongside your test plates.

5. Make sure you swirl the plates gently against the bench to assure even mixing, within a few minutes after adding the agar. If you mix too roughly, you can splash agar onto the top of the dish, which can reduce the accuracy of the results.

6. Rinse the agar bottles with water, and place them on a designated cart. (This will help us with dish washing later.)

7. Leave pour plates on the bench until solidified.

8. After the agar solidifies, invert the plates and incubate at 32°C for 48 h.

VRBA Spread Plate
Procedure

1. Label the VRBA plates (group names, plated dilution). We will be plating each dilution in duplicate.

2. Plate dilutions 10^{-2} through 10^{-5}.

3. Use a flamed "hockey stick" to spread the liquid.

4. After the liquid soaks into all the plates, add enough tempered VRBA to cover the top of the agar surface of each plate.

5. After the overlay hardens, invert the plates and incubate at 32°C for 48 h.

Class 2

Count the SPC colonies.

Count the typical coliform colonies (purplish red with bile precipitate).

Use long-wave UV light to count MUG + *E. coli*.

■■ IV. RESULTS

Standard Plate Count

Dilution as Plated	CFU/Plate	CFU/Plate	Average CFU/Plate
10^{-3}			
10^{-4}			
10^{-5}			
10^{-6}			

SPC CFU/g =

VRBA Presumptive Coliform Count

Dilution as Plated	CFU/Plate	CFU/Plate	Average CFU/Plate
10^{-2}			
10^{-3}			
10^{-4}			
10^{-5}			

Presumptive Coliform CFU/g =

VRBA Presumptive *E. coli* Count

Dilution as Plated	CFU/Plate	CFU/Plate	Average CFU/Plate
10^{-2}			
10^{-3}			
10^{-4}			
10^{-5}			

Presumptive *E. coli* CFU/g =

■ V. DISCUSSION QUESTIONS

1. Assume countable plates with 25 to 250 CFU/plate. What is the minimum and maximum CFU/g you can accurately detect with the SPC dilution scheme? What about the minimum and maximum coliform counts?

2. If the SPC plates were incubated at 37°C, would you expect different results?

3. Why is your coliform count considered to be presumptive? What other experimental parameters do you need to test to confirm coliforms?

■ LABORATORY NOTES

■ LABORATORY NOTES

▬ LABORATORY NOTES

LABORATORY 6

ESCHERICHIA COLI O157: H7 ENRICHMENT AND IMMUNOMAGNETIC SEPARATION

I. OBJECTIVES

- Understand the use of enrichments in the isolation of pathogens from foods.

- Learn the basics of immunomagnetic separation.

II. BACKGROUND

The detection of *E. coli* O157:H7 in food products involves selective enrichment. Selective enrichment is performed to encourage growth of your target pathogen and inhibit growth of background organisms. Selective enrichment is used when the level of the organism is not a priority, instead, the researcher is interested in its presence or absence. Enrichments are usually used for serious human pathogens, such as *Salmonella*, *E. coli* O157:H7, and *Listeria monocytogenes*. Enrichments generally use samples of food products (often 25 g), which are homogenized in a selective enrichment broth. This food homogenate is incubated for 18 to 24 h to allow levels of the pathogen of interest to increase and multiply, while inhibiting other organisms. Theoretically, after selective enrichment, your pathogen had time to grow to very large numbers, even if it was in your original food product at very low levels. It is much easier to isolate and identify your pathogen of choice from a selective enrichment broth due to inhibition of other microflora (due to selective agents) and increased numbers of target organisms.

The current official U.S. Department of Agriculture (USDA) method[10] for enrichment and detection of *E. coli* O157:H7 involves testing a total of 325 g of ground beef. In routine ground beef testing programs, five 65 g subsamples are each homogenized in 585 ml modified *E. coli* enrichment broth with novabiocin (mEC+n). However, for outbreak-related samples, thirteen 25 g subsamples are each homogenized into 225 ml mEC+n. After enrichment, cells are concentrated from 1 ml of enrichment broth using immunomagnetic separation with magnetized beads covered with anti-O157 antibodies. This step increases the sensitivity through concentration and the specificity of the method by physically separating *E. coli* O157 cells from much (but not all) of the background microflora. The beads are then plated on Rainbow® Agar O157 (BiOLOG, Hayward, CA). Typical O157 colonies on this media (gray–black color) are then confirmed using biochemical testing and serological confirmation.

Media Used in this Lab

Modified E. coli Broth with Novobiocin

This broth is designed for mEC selective enrichment of *E. coli* O157:H7 and other enterohemorrhagic *E. coli* (EHEC) in foods. This media contains peptone and salt (general nutrients), lactose is added to improve growth of lactose-positive organisms, phosphate is added to buffer the medium, and a mixture of bile salts number 3 and novobiocin suppresses the growth of Gram-positive microflora.

Sorbitol–MacConkey Agar (SMAC) with MUG

This media contains peptone (nitrogen source), sorbitol (sugar), bile salts, and crystal violet (to inhibit growth of Gram-positive organisms), NaCl, neutral red (pH indicator), and 4-methylumbel-liferyl-β-D-glucuronide (MUG) (differential agent). *E. coli* O157:H7 differs metabolically from the majority of *E. coli* in that it cannot utilize the sugar sorbitol, and it does not produce the enzyme β-D-glucuronidase. Therefore, on this media, the majority of sorbitol-positive *E. coli* will appear pink-red in color and will be fluorescent under UV light. *E. coli* O157:H7 will appear as a nonfluorescent white colony. One limitation to this media is that other organisms (such as a member of the *Proteus* genus) will appear as colorless, nonfluorescent colonies on this media; therefore, colonies must undergo further confirmation using biochemical and serological tests.

Rainbow Agar O157

Rainbow Agar O157 (BiOLOG, Hayward, CA) is both selective and differential. To increase selectivity, the addition of 0.8 mg/liter potassium tellurite and 10 mg/liter novobiocin can be added to the manufacturer's formulation. The ingredients in the Rainbow Agar are proprietary; however, the differential nature uses chromogenic substrates directed against two *E. coli* enzymes: one substrate turns blue-black when colonies have β-galactosidase activity, and the second substrate turns red when colonies have β-glucuronidase activity. The majority of *E. coli* O157:H7 are β-glucuronidase negative. Therefore, the colony color is blue-black, reflecting only the β-galactosidase activity. Other entero-toxigenic serotypes and nonpathogenic *E. coli* colonies will range in color from pink or magenta to purple or blue, depending on the relative levels of each enzyme produced. Background organisms tend to be white to cream in color. Other members of *Enterobacteriaceae* (*Klebsiella pneumoniae* and sometimes *Hafnia alvei* or *Citrobacter*) look similar to *E. coli* O157:H7. All suspicious colonies should be further identified using biochemical and serological tests.

■ III. METHODS

Note: **In this laboratory, you may isolate a serious human pathogen. Therefore, each step should be treated as if that pathogen is present. As a precaution, you should wear eye protection and gloves during all steps.**

This procedure is an adaptation of the current official method of the USDA.[10] A flowchart of this lab is presented in Figure 6.1.

Class 1

Sample Preparation and Enrichment

Procedure

1. Weigh a 25 g sample of ground beef into a filter Stomacher® bag. Add 225 ml of mEC+n broth. Homogenize for 2 min in a Stomacher blender.
2. Loosely close bags and place in a 1 liter beaker to stabilize. Incubate at 35 to 37°C for 20 to 24 h.

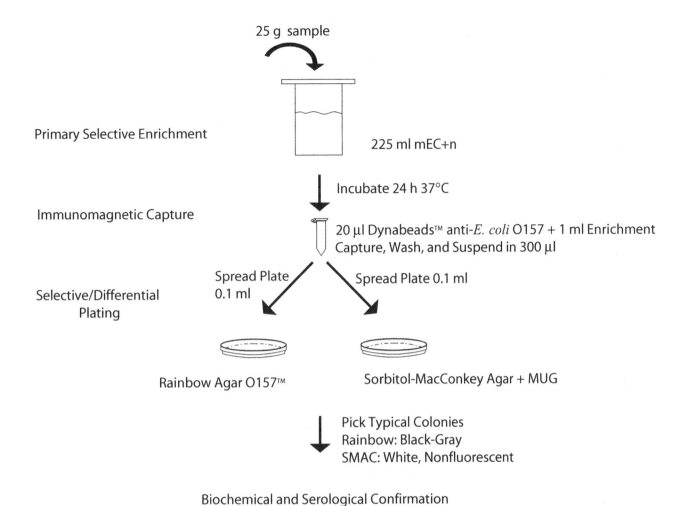

25 g sample

Primary Selective Enrichment

225 ml mEC+n

Incubate 24 h 37°C

Immunomagnetic Capture

20 µl Dynabeads™ anti-*E. coli* O157 + 1 ml Enrichment
Capture, Wash, and Suspend in 300 µl

Spread Plate
0.1 ml

Spread Plate 0.1 ml

Selective/Differential
Plating

Rainbow Agar O157™

Sorbitol-MacConkey Agar + MUG

Pick Typical Colonies
Rainbow: Black-Gray
SMAC: White, Nonfluorescent

Biochemical and Serological Confirmation

Figure 6.1   The enrichment and immunocapture of *E. coli* O157:H7.

Class 2

Magnetic Immunocapture and Plating

Procedure

1. Label a 1.5 ml microcentrifuge tube for each enrichment.
2. Thoroughly resuspend the Dynabeads® anti-*E. coli* O157 (Dynal Inc. Lake Success, NY) by vortexing until the pellet in the bottom disappears.
3. **Pipette 20 µl of Dynabeads anti-*E. coli* O157 into each tube.**
4. Add 1 ml of the preenriched sample to each tube. Be sure to remove sample from the outside of the filtered inner bags. Close each tube tightly.
5. Invert the tubes three times to mix the Dynabeads and the samples. Incubate 10 min on a rotary mixer, allowing for gentle continuous agitation to prevent the beads from settling. During this time, cells with the appropriate surface structures (O serotype 157) will be bound to the magnetic beads.
6. Place the tubes in a magnetic capture rack without the magnetic plate. Once all the tubes are in place, move the magnetic plate back into the rack. Invert the rack several times to concentrate the beads in each tube into a pellet on the side of the tube. Allow 3 min for proper recovery. If present, the *E. coli* O157 cells will be concentrated in this pellet in each tube.

7. Leave the magnetic plate in place. In a biological safety hood (if available), carefully open the tubes. Remove and discard the sample supernatant and any liquid remaining in the caps of the tubes.

8. Remove the magnetic plate. Add 1 ml sterile wash buffer (PBS-Tween). Close each tube firmly and invert to suspend the beads.

9. Repeat steps 6 through 8 twice.

10. After the third washing step, remove the magnetic plate and suspend the beads in 300 μl of wash buffer (PBS-Tween).

11. Transfer 100 μl suspended beads to dry plates (no surface moisture) of Rainbow Agar O157 and Sorbitol-MacConkey Agar with MUG. Spread the beads using a sterile "hockey stick." Plates can be dried ahead of time by opening them in a laminar flow hood for up to 30 min.

12. Incubate for 24 to 26 h at 35 to 37°C.

Class 3

Results of Immunocapture and Plating

1. Study plates for typical *E. coli* O157:H7 colonies. On Rainbow Agar O157, typical *E. coli* O157:H7 colonies will be blue-black in color.

2. On Sorbitol-MacConkey Agar with MUG, *E. coli* O157:H7 colonies will appear colorless and will not fluoresce under UV light.

3. If typical colonies are present, they are presumptive *E. coli* O157. To confirm they are O157:H7, a number of other tests must be performed:

   a. Biochemical confirmation: This confirms that the isolated colonies are *E. coli*. A miniaturized biochemical detection unit (such as the API® 20E [bioMérieux, Raleigh, NC] or Enterotube™ II [Becton, Dickinson and Company, Franklin Lakes, NJ]) can be used. Colonies from beads should be streaked to confirm purity on a nonselective media before performing these tests.

   b. Serological confirmation: The presence or absence of the O157 (lipopolysaccharide O) and H7 (flagellar) antigens can be performed using a rapid latex agglutination test kit (such as the RIM® *E. coli* latex Test Kit [REMEL, Lenexa, KS]).

   Shiga toxin production: The presence of Shiga toxin should be confirmed using a toxin assay (such as the Premier™ EHEC kit from Meridian Bioscience, Inc., Cincinnati, OH).

▬ IV. RESULTS

Describe colony morphologies on each of the media. Were any presumptive *E. coli* O157:H7 present?

▬ V. DISCUSSION QUESTIONS

1. What is the advantage to using immunocapture prior to selective differential plating?

2. Why are the colonies from the SMAC and Rainbow Agar presumptive *E. coli* O157?

3. The Food Safety and Inspection Service of the U.S. Department of Agriculture typically tests 325 g of ground beef. Discuss the differences between the routine-testing scheme (five enrichments of 65 g) and the outbreak-testing scheme (thirteen enrichments of 25 g)?

▰ LABORATORY NOTES

_____

_____

_____

_____

_____

_____

_____

_____

_____

_____

_____

_____

_____

_____

_____

_____

_____

_____

LABORATORY NOTES

LABORATORY 7

DETECTION AND IDENTIFICATION OF *SALMONELLA* SPP.

I. OBJECTIVES

- Become familiar with the different steps involved in isolating and identifying *Salmonella* spp. in foods.

- Become familiar with the use of selective enrichment for isolating pathogens.

II. BACKGROUND

Salmonella is a member of the *Enterobacteriaceae* family and is always considered a pathogen when isolated from food products or humans. The disease caused by *Salmonella* is generically called salmonellosis and can range from bacterial diarrhea to septicemia. Bacterial diarrhea salmonellosis infection is generally foodborne, usually reaching humans through contaminated eggs or improperly cooked poultry products. Fowl harbor many species of *Salmonella* in their gastrointestinal (GI) tracts, which can result in contamination if handled improperly. *S. typhi*, the cause of typhoid fever, differs from most *Salmonella* spp. in that its reservoir is human rather than animal. The organism is spread from person to person via the oral–fecal route.

The taxonomy of the *Salmonella* genus is complex and is based on both biochemical and serological properties. We will not have time in this lab exercise to classify *Salmonella* through serology. *Salmonella enterica*, which is divided into five subspecies, contains most of the human disease agents and is comprised of over 2000 serotypes based on their serologic properties. The unique serotype components are the flagellar protein antigens (H), the cell wall polysaccharide antigens (O), and the capsular polysaccharide antigen (Vi).

The methods for isolating and identifying *Salmonella*, as outlined in the *Bacteriological Analytical Manual* (BAM) published by the U.S. Food and Drug Administration (FDA), consist of five steps: preenrichment, selective enrichment, selective plating, metabolic testing, and serology.[11] These are summarized in Figure 7.1. We will do the first four steps of this process using a number of different samples and sample preparation techniques. In the metabolic testing steps, there are more than 20 different possible tests, but we will be conducting a small series of metabolic tests consisting of triple sugar iron agar, lysine iron agar, and the IMViC tests. The IMViC series consists of indole, methyl red, Voges–Proskauer, and citrate tests.

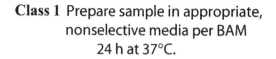

Class 1 Prepare sample in appropriate,
nonselective media per BAM
24 h at 37°C.

Class 2 Transfer into enrichment broths. Inc. 24 h at 37°C.

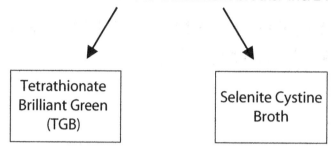

| Tetrathionate Brilliant Green (TGB) | | Selenite Cystine Broth |

Class 3 Streak from enrichments onto each selective/differential media.

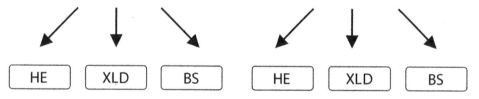

| HE | XLD | BS | | HE | XLD | BS |

Class 4 Pick 2 characteristic Salmonella colonies* from each plate and
inoculate one TSI and LIA slant for each colony.
*HE: Blue-green colonies with or without black center
XLD: Pink colonies with or without black centers
BS: Brown, gray, or black colonies, sometimes with a metallic sheen

Class 5 Observe TSI and LIA slants. Inoculate Tryptone broth,
MRVP broth and Simmons citrate agar from + TSI slants.

Class 6 Perform Indole, Methyl Red, and Voges-Proskauer tests.
Observe Simmons citrate result. Tabulate results.

Figure 7.1 Flowchart of traditional isolation of *Salmonella* spp. from foods.

Bacteriological Media Used in this Lab

For more information, see the *Bacteriological Analytical Manual* published by the U.S. Food and Drug Administration.[11]

Nonselective Enrichment

1. Tryptic soy broth (TSB): This is a general-purpose nutrient media used for cultivating fastidious and nonfastidious microorganisms. The medium is a soybean-casein digest.

2. Lactose broth (LB): This medium is often used as a preenrichment broth for coliform bacteria and other members of the *Enterobacteriaceae* family.

Selective Enrichment Broths

3. Selenite cystine (SC) broth: This is a selective enrichment broth recommended for use in detecting and identifying *Salmonella* in dairy products and other foods. Sodium selenite is used as the selective agent against coliforms. Cystine is added to facilitate the growth of *Salmonella*.

4. Tetrathionate (TT) broth: This is a selective enrichment broth for detecting *Salmonella* in foods. This media contains proteose peptone (a peptic digest of animal tissues), bile salts, sodium thiosulfate, and calcium carbonate. During media preparation, an iodine solution is added. The iodine reacts with thiosulfate, leading to TT formation. TT inhibits the growth of coliform bacteria, while it allows for the growth of *Salmonella*. Bile salts reduce the growth of Gram-positive organisms. This media also has calcium carbonate as a buffer, and general nutrients are supplied by proteose peptone.

Selective/Differential Media

5. Bismuth sulfite (BS) agar: This is a selective medium for the isolation of *Salmonella typhi* and other *Salmonella* spp. BS and brilliant green are used as the selective agents against growth of Gram-positive bacteria, while allowing *Salmonella* to grow. Disodium phosphate is added to buffer the media; dextrose is the energy source; and beef extract and peptone supply general nutrients. Ferrous sulfite is added for hydrogen sulfide production. *Salmonella* colonies are generally black or greenish-grey, with or without a metallic sheen.

6. Xylose desoxycholate (XLD) agar: This medium is a selective differential plating medium designed for direct isolation of *Shigella* and *Providencia* from stool samples. Sodium desoxycholate is used to inhibit the growth of Gram-positive bacteria, and it inhibits the swarming of *Proteus*. Salmonella is differentiated from other bacteria by using three reactions: xylose fermentation, lysine decarboxylation, and hydrogen sulfide production. *Salmonella* rapidly ferments xylose, thus stimulating the decarboxylation of lysine, producing alkaline conditions (red colony). This alkaline environment allows for the hydrogen sulfide production, causing black centers in *Salmonella* colonies while other enteric organism colonies remain acid yellow or red with no black center.

7. Hektoen enteric (HE) agar: This is a differential and selective medium used for isolating and differentiating *Salmonella* and *Shigella* genera from other Gram-negative enteric organisms. This medium has raised carbohydrate and peptone contents to counteract the inhibitory effects of the bile salts and indicators. This combination only slightly inhibits the growth of *Salmonella* and *Shigella*, while inhibiting other enteric organisms. The lactose-positive fermenting organisms are differentiated from lactose-negative colonies due to the presence of two indicators, bromthymol blue and acid fuchsin. Finally, the combination of thiosulfate and ammonium citrate causes hydrogen sulfide-producing colonies to become black. Lactose-positive colonies appear as yellow to orange colonies. Bacteria that produce hydrogen sulfide appear as colonies with black centers.

Metabolic Testing

8. Triple sugar iron (TSI) agar: This medium is recommended for identifying Gram-negative enteric bacilli based on the fermentation of glucose, lactose, and sucrose and for hydrogen sulfide production. Demonstration of the organism's fermentative properties depends on the proportions of the carbohydrates (10 parts lactose:10 parts sucrose:1 part glucose), as well as their concentrations relative to the peptone concentration. Phenol red serves as the acid indicator, and ferrous sulfate is added for hydrogen sulfide formation. After autoclaving, the medium is solidified at an angle to create a slanting top on a solid butt. When glucose is fermented, the entire medium becomes yellow (acid) in 8 to 12 h. The butt will remain acidic after 18 to 24 h due to the presence of

organic acids resulting from anaerobic conditions. The slant, however, will revert to the alkaline (red) color due to the oxidation of peptones under aerobic conditions on the slant, with the formation of carbon dioxide, water, and alkaline amines. When lactose or sucrose is fermented in addition to glucose, the large amounts of acid products more than neutralize the alkaline products on the slant, causing the slant to remain yellow in color. Formation of carbon dioxide and hydrogen is indicated by cracks or bubbles in the agar. The production of hydrogen sulfide requires an acidic environment and is shown by blackening of the butt. The reaction must be read between 18 to 24 h of incubation, because the parameters set by the media formulation will not last over time.

   Alkaline slant and alkaline butt (K/K) = nonfermenter

   Alkaline slant and acid butt (K/A) = glucose fermentation only

   Acid slant and acid butt (A/A) = glucose, sucrose, or lactose fermenter

   K/A/gas + H_2S = glucose fermentation with gas production and H_2S production

9. Lysine iron agar (LIA): This medium is designed to determine whether a bacterium decarboxylates or deaminates lysine and forms hydrogen sulfide. In an acidic environment, if the organism produces a coenzyme pyridoxal phosphate and an enzyme decarboxylase, amino acids are decarboxylated to the corresponding amine. The medium contains peptones, glucose, ferric ammonium citrate, and sodium thiosulfate, in addition to lysine, and is made into an agar slant. When glucose is fermented, the butt turns yellow in color. If lysine is decarboxylated, a compound called cadaverine is produced, typical of *Salmonella*. Cadaverine neutralizes the organic acids formed from glucose fermentation, and the medium is reverted to the purple color. The protea genus of the *Enterobacteriaceae* family can carry out an oxidative deamination of lysine yielding a compound which, in the presence of ferric ammonium citrate and a coenzyme (flavin mononucleotide), forms a burgundy red color.

   Alkaline slant and acid butt (K/A) = glucose fermentation

   Alkaline slant and alkaline butt (K/K) = lysine decarboxylation or no fermentation

   Red slant and acid butt (R/A) = lysine deamination and glucose fermentation

10. Tryptone broth: This medium is a pancreatic digest of casein and is used in this lab because of its high tryptophane content, in differentiating cultures that produce or do not produce indole.

11. MR-VP (methyl-red Voges–Proskauer) broth: This is a simple broth medium composed of glucose, peptone, and some salts. It is used to discriminate between organisms using a mixed acid pathway (strong acids) of metabolism (detected by the methyl red test) or a butylene glycol pathway generating acetoin, detected in the Voges–Proskauer test.

12. Simmons citrate agar: This medium is used to determine if an organism can use sodium citrate as its sole source of carbon, which would mean that it is a differential medium. If sodium citrate is utilized, alkaline products including NaOH are produced. The indicator used is bromothymol blue, which is blue at alkaline pH.

■ III. METHODS

NOTE: **In this laboratory, you may isolate a serious human pathogen. Therefore, each step should be treated as if that pathogen is present. As a precaution, you should wear eye protection and gloves during all steps.**

Class 1

Sample Preparation and Preenrichment
These enrichments are adapted from the *Bacteriological Analytical Manual* (U.S. Food and Drug Administration).[11] Each group will be given one of the foods listed below. Chicken can be prepared for enrichment using the blender or Stomacher methods.

Procedure

>   Shell eggs: Soak the eggs in 70% ethanol for 30 min. Crack the eggs aseptically, and, using a sterile egg separator, discard the whites. Aseptically weigh 25 g of egg yolk into a sterile 500 ml screw cap jar. Add 225 ml TSB, and mix well by swirling. Loosen cap ¼ turn and incubate for 24 h at 37°C.
>
>   Chicken (blender method): Aseptically remove a 25 g sample of breast skin from the whole bird using sterile scissors and forceps. Place the sample and 225 ml lactose broth in a sterile blender jar, and blend for 2 min. Aseptically transfer the mixture to a sterile 500 ml screw cap jar. Loosen cap a ¼ turn, and incubate the jar at 37°C for 24 h.
>
>   Chicken (Stomacher method): Remove the chicken skin as above and add to a Stomacher bag with 225 ml lactose broth. Homogenize in a paddle blender (such as a Stomacher) for 5 min on the highest setting. Seal the bag loosely, and incubate the bag in a 600 ml beaker at 37°C for 24 h.

Class 2

Selective Enrichment

NOTE: **In this laboratory, you may isolate a serious human pathogen. As a precaution, you should wear eye protection and gloves during all steps after the initial homogenization of the product in enrichment broth.**

Procedure

Gently mix your enrichments, and aseptically transfer 1 ml volumes to two 10 ml SC and TT broth tubes. You will have a total of four selective enrichment tubes (two of each medium) from each preenrichment. Incubate at 37°C for 24 h.

Class 3

Selective Plating

Procedure

1. Carefully mix the turbid selective enrichments using a vortex mixer. If the tubes have Morton's closures (looks similar to an upside-down bell), be careful not to vortex so vigorously that the sample spills from the tube.
2. Use an inoculation loop to streak on BS agar, XLD agar, and HE agar. All three media should be streaked from each selective enrichment tube.
3. Incubate at 35°C for 24 h.

NOTE: **At this point, you will have 12 plates generated from a single food sample. From here on, you have to be organized in order to keep track of your possible *Salmonella* cultures. Table 7.1 is provided to record results as you work through this lab. You should develop a labeling scheme and keep a key in your lab notebook.**

Class 4

Plate Examination and Inoculation of Metabolic Test Media

1. Examine plates for the presence of suspicious *Salmonella* colonies. Positive streak examples will be provided for comparison:
    a. HE: Typical colonies are blue-green to blue, with or without black centers. Often, *Salmonella* colonies will be almost completely black and glossy.

    b. BS: Typical *Salmonella* colonies may appear brown, gray, or black and may sometimes have a metallic sheen. The surrounding medium will start out as brown, turning black over the incubation time. Some strains produce green colonies with little or no darkening of the surrounding medium.

    c. XLD: Typical colonies will be pink, with or without black centers. *Salmonella* colonies will often be almost completely black and glossy.

2. Select two suspected *Salmonella* colonies from each selective agar, recording its colony morphology. Inoculate each colony into TSI agar and LIA. Incubate one tube of uninoculated media as a negative control. This will help determine if a change occurred. Do not discard the plates. Wrap them with parafilm and keep them in case of further analysis.

3. Inoculate a TSI slant by lightly touching a suspicious colony with a sterile inoculating needle and streaking the slant in a zigzag motion and ending with a stab of the butt of the slant.

4. Inoculate an LIA slant by touching the same colony with a sterile needle as above and then stabbing the butt twice and finally streaking the slant as above.

5. Incubate TSI and LIA slants at 37°C for 24 h.

Class 5

Further Metabolic Testing

1. Examine TSI slants. A typical *Salmonella* culture produces an alkaline (red) slant and an acid (yellow) butt, with or without H_2S production (blackening of agar). Record the results in Table 7.1.

2. Examine LIA slants for an alkaline (purple) reaction in the butt of the tube, typical of *Salmonella* cultures. Be careful to consider only a distinct yellow color in the butt of the tube as a negative acid reaction, because some *Salmonella* cultures may produce a slight discoloration in the butt. Most *Salmonella* cultures produce H_2S in LIA. Record the results in Table 7.1.

3. With positive TSI slants, inoculate the tubes of tryptone broth. This is done using a sterile inoculating loop. Incubate at 35°C for 24 h. This tube will be used to carry out the indole test.

4. Inoculate MR-VP broth tubes with positive TSI slants. Use the same procedure as above in #3, and this culture tube will later be used for the methyl red and Voges–Proskauer tests. Incubate at 35°C for 48 h.

5. Finally, inoculate Simmons citrate agar slants with positive TSI slants by streaking the slant and stabbing the butt using a sterile needle. Incubate at 35°C for 46 h.

Class 6

Metabolic Test Results

Perform the following tests, and record the results in Table 7.1:

1. Indole test: Add three to five drops of Kovac's reagent to your tryptone broth tubes using Pasteur pipettes. Most *Salmonella* cultures give a negative test (no red color at the surface of the broth).

2. Voges–Proskauer (VP) test: Transfer 1 ml culture from MR-VP broth to a sterile test tube, add 0.6 ml -naphthol, and shake well. Add 0.2 ml 40% KOH solution and shake. Read the results after 4 h: development of a pink to ruby red color throughout the broth shows a positive test. *Salmonella* cultures are typically VP negative.

3. Methyl red test: Transfer 5 ml of MR-VP broth to a sterile test tube, and add five to six drops of methyl red indicator. Results are read immediately. A positive test is confirmed by a red color throughout the tube. *Salmonella* cultures are typically methyl red positive.

4. Examine Simmons citrate agar: A positive result is indicated by a color change from green to blue. Most cultures of *Salmonella* are citrate positive. Negative results are shown by little or no growth with no color change.

IV. RESULTS

TABLE 7.1
Tabulation of *Salmonella* Laboratory Results

Selective Enrichment	Selective Plating Media	Your Own Strain ID Code	TSI Agar	LIA	Indole	MR	VP	Citrate	Final Identification
Selenite cystine (SC-1)	BS								
	XLD								
	HE								
Selenite cystine (SC-2)	BS								
	XLD								
	HE								
Tetrathionate broth (TT-1)	BS								
	XLD								
	HE								
Tetrathionate broth (TT-2)	BS								
	XLD								
	HE								

■ V. DISCUSSION QUESTIONS

1. Justify the use of the preenrichment step, rather than simply placing your sample directly into the selective enrichment broth, knowing that the preenrichment medium is a nutrient medium allowing a wide variety of organisms to grow.

2. What would be the advantage in doing even more biochemical tests than the number we did in lab?

3. Discuss how this long *Salmonella* identification process could impact a food-processing plant's product distribution practices?

4. Why is it important to start the biochemical testing step with a single well-isolated colony from a selective agar plate?

5. Did you have difficulty interpreting the results of the different tests? Could the experience levels of lab workers impact the reliability of quality control testing? Explain.

6. What are some new technologies that could shorten or eliminate this long process? Where in the process could these be implemented?

TABLE 7.2[12]

Biochemical Reactions of Selected Genera of Enterobacteriaceae[a]

Organism	TSI[b] (slant/butt)	LIA[c] (slant/butt)	H$_2$S[d]	Gas[d]	Indole[e]	MR[f]	VP[g]	Citrate[h]
Citrobacter	A/A	K/A	+/−	+	+/−	+	−	+
E. coli	A/A	K/K	−	+	+	+	−	−
Edwardsiella	K/A	K/K	+/−	+	+	+	−	−
Enterobacter	A/A	K/A	−	+	−	−	+	+
Klebsiella	A/A	K/K	−	+	−	−	+	+
Morganella	K/A	R/A	−	+	+	+	−	−
Proteus	A/A	R/A	+/−	+	+/−	+	−	+/−
Providencia	K/A	R/A	−	−	+	+	−	+
Salmonella	K/A	K/K	+	+	−	+	−	+
Serratia	A/A	K/A	−	+	−	−	+	−
Shigella	K/A	K/A	−	−	+/−	+	−	−

[a] Basic key: + the majority of organisms in this genus have this characteristic; − the majority of organisms in this genus lack this characteristic; +/− this characteristic varies by species within this genus.

[b] K = alkaline (red); A = acid (yellow or black [acid + H$_2$S]).

[c] K = alkaline (purple indicates no fermentation or lysine decarboxylation); A = acid (yellow); R = deamination reaction (burgundy color).

[d] Hydrogen sulfide (black color) is produced from ferrous sulfate produced in an acidic environment. Gas (CO$_2$ and H$_2$) is indicated by cracks or separations in the agar. Both of these reactions can be observed with TSI and LIA slants.

[e] + = a red color observed at the top of the broth when Kovac's reagent is added to growth in tryptone broth; − = a colorless ring.

[f] Methyl Red. + = a red color reaction throughout the tube with methyl red indicator showing a mixed acid metabolism in MR-VP broth growth; − = no color change.

[g] Voges–Proskauer. + = a pink to ruby red color throughout the tube with the addition of -naphthol and KOH indicating acetoin production; − = no color change occurs.

[h] + = color change from green to blue; − = no color change (green).

Source: Adapted from Holt, J.G., Krieg, N.R., Sneath, P.H.A., Staley, J.T., and Williams, S.T., *Bergey's Manual of Determinative Bacteriology,* 9th ed., Williams and Wilkins, Baltimore, 1994.

■ LABORATORY NOTES

■ LABORATORY NOTES

LABORATORY 8

ENRICHMENT MPN OF *VIBRIO PARAHAEMOLYTICUS* FROM SHRIMP

▂▂ I. OBJECTIVES

- Learn to use an enrichment MPN for estimating pathogen levels.

- Become familiar with methods used to isolate *Vibrio* spp.

▂▂ II. BACKGROUND

The *Vibrio* genus consists of several species. *V. cholerae*, *V. vulnificus*, and *V. parahaemolyticus* are significant with respect to foods and human health.[13] *Vibrios* are naturally present in marine environments. *V. parahaemolyticus* is usually associated with coastal and estuary waters and can be found in fish, shellfish, and crustaceans harvested from these waters. Environmental isolation of *V. parahaemolyticus* is dependent upon water temperature: when the temperature falls below 15°C, it cannot be isolated from water; however, survival during winter is thought to be due to survival in sediments.[14]

With respect to cellular morphology, *V. parahaemolyticus* cells are facultative Gram-negative short to comma-shaped rods possessing motility by a single polar flagellum. They are metabolically both oxidative and fermentative and most importantly obligate halophiles (salt loving), which is used to separate *V. parahaemolyticus* from *V. cholerae*.[13]

V. parahaemolyticus is the leading cause of foodborne illness in Japan and has caused outbreaks in the United States. The most common syndrome is watery diarrhea that affects patients 4 to 96 h after consumption and lasts about 3 days. However, a high ingestion of greater than 10^5 cells can produce dysentery syndrome. Symptoms include mucus and blood in stools, affecting patients just 20 min to 9 h after ingestion and lasting about 2.5 days. This syndrome usually occurs in summer to early fall. The mechanism of pathogenecity is not well known, but most clinical isolates exhibit a thermostable hemolysin toxin responsible for the Kanagawa phenomenon.[13]

The enrichment, isolation, and identification of members of the genera *Vibrio* are based upon their ability to grow well at alkaline pH. *V. parahaemolyticus* is further characterized based upon its requirement for NaCl for growth.

Bacteriological Media Used in this Lab

1. Alkaline peptone water (APW): This enrichment contains 1% peptone and 3% NaCl per liter of water. The pH is adjusted to 8.5. This enrichment is selective based upon the raised pH and the high level of NaCl.[15]

2. Thiosulfate-citrate-bile salts-sucrose (TCBS) agar. This is a selective differential medium. Nutrients are provided by yeast extract, peptone, vitamins and amino acids. Selective agents include sodium citrate, sodium thiosulfate and oxgall, which provide an alkaline pH (final pH approx 8.6) that slows growth coliforms and Gram positive organisms. In addition, sodium thiosulfate and ferric citrate act together as an indicator to detect H_2S production (blackening a media). Sucrose (or saccharose) is added as the fermentable carbohydrate to detect sucrose fermentation with brom thymol blue and thymol blue as the pH indicators. All pathogenic members of *Vibrio* will grow on this media with the exception of *V. hollisae*.[11]

3. Tryptic soy agar (TSA) + NaCl: In order for halophilic *Vibrio* spp. to grow in this medium, 1.5% NaCl must be added to a final concentration of 2%. This is a nondifferential and nonselective media.

4. Gelatin agar (GA): This media contains peptone (0.4%), yeast extract (0.1%), gelatin (1.5%), and agar (1.5%), with a final pH at 7.2. The selectivity of this media is based upon the lack of NaCl, because halophilic *Vibrio* spp. require NaCl for growth.

5. Gelatin salt agar (GSA): This medium has the same composition as GA, with the addition of 3% NaCl.

6. Triple sugar iron (TSI) agar + NaCl: In order for halophilic *Vibrio* spp. to grow in this medium, 1.5% NaCl must be added to the final level of 2%. This medium is used to differentiate carbohydrate utilization and has a specific proportion of carbohydrates (10 parts lactose:10 parts sucrose:1 part glucose). Phenol red serves as the acid indicator, and ferrous sulfate is added for hydrogen sulfide formation. After autoclaving, the medium is solidified at an angle to create a slanting top on a solid butt. When glucose is fermented, the entire medium becomes yellow (acid) in 8 to 12 h. The butt will remain acidic after 18 to 24 h due to the presence of organic acids resulting from anaerobic conditions. The slant, however, will revert to the alkaline (red) color due to the oxidation of peptones under aerobic conditions on the slant with the formation of carbon dioxide, water, and alkaline amines. When lactose or sucrose is fermented in addition to glucose, the large amounts of acid neutralize any alkaline products and allow the slant to remain yellow in color. Formations of carbon dioxide and hydrogen gases are indicated by cracks or bubbles in the agar. The production of hydrogen sulfide requires an acidic environment and is shown by a blackening of the butt. The reaction must be read between 18 to 24 h of incubation, because the parameters set by the media formulation will not last over time:

    Alkaline slant and alkaline butt (K/K) = nonfermenter

    Alkaline slant and acid butt (K/A) = glucose fermentation only

    Acid slant and acid butt (A/A) = glucose, sucrose, or lactose fermenter

7. Motility test medium (MTM) supplemented + NaCl: In order for halophilic *Vibrio* spp. to grow in this medium, 1.5% NaCl must be added to the final concentration of 2%. This is a nonselective, soft agar (0.5% agar) that is prepared in test tubes. Each tube is inoculated by stabbing ¾ of the depth. Motility is defined as a diffuse zone of spreading growth from the line of inoculation.

▪ III. METHODS

NOTE: **In this laboratory, you may isolate a serious human pathogen. Therefore, each step should be treated as if that pathogen is present. As a precaution, you should wear eye protection and gloves during all steps.**

This laboratory will take place over five class periods. The steps are outlined in Figure 8.1.

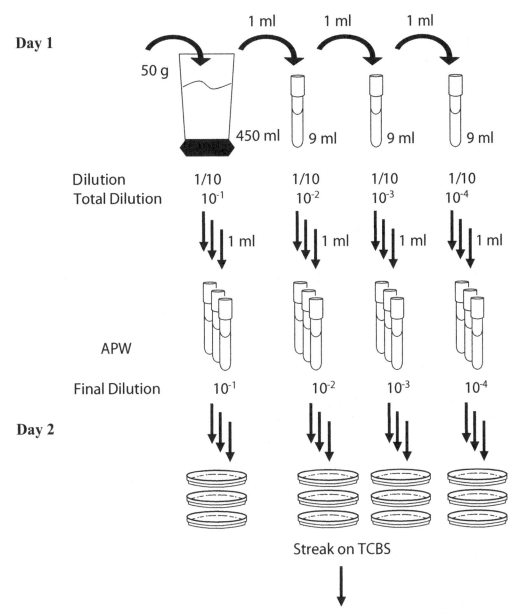

Day 1

1 ml   1 ml   1 ml

50 g

450 ml   9 ml   9 ml   9 ml

| Dilution | 1/10 | 1/10 | 1/10 | 1/10 |
| Total Dilution | 10^{-1} | 10^{-2} | 10^{-3} | 10^{-4} |

1 ml   1 ml   1 ml   1 ml

APW

Final Dilution 10^{-1} 10^{-2} 10^{-3} 10^{-4}

Day 2

Streak on TCBS

Day 3 Record dilution and number of plates with typical *V. parahaemolyticus* colonies.
Pick 2 round, 2 to 3 mm diameter green or blue-green colonies from each plate,
and patch on GSA and GA and streak TSA+NaCl.

Day 4 Use gelatinase + colonies, with growth on GSA but not GA,
for Gram stain, oxidase test and inoculate TSI+NaCl, MTM+NaCl
and TSA+NaCl for O/129 sensitivity testing.

Figure 8.1 Flowchart of enrichment MPN to determine levels of *V. parahaemolyticus* from shrimp.

Class 1

Sample Preparation and Enrichment MPN

Procedure

1. Measure 50 g of shrimp (including shell and gut) into a sterile tarred blender jar. Add 450 ml sterile 2% NaCl solution, and blend for 2 min at high speed. This is the first 1:10 dilution.
2. Prepare a dilution series (Figure 8.1) in 9 ml 2% NaCl tubes.
3. Remove 1 ml portions from dilutions, and inoculate 10 ml APW.
4. Incubate 16 to 18 h at 35 to 37°C.

Class 2

Streak onto TCBS from Each Enrichment MPN

Procedure

1. All tubes that are turbid and at least one dilution higher should be streaked onto TCBS agar. It is important not to shake the tubes. Streak the agar with a loopful from the top 1 cm of each enrichment broth. Be sure to label plates with dilution and tube number.

Class 3

1. Examine TCBS agar for typical *V. parahaemolyticus* colonies. They will appear round, 2 to 3 mm in diameter, and green or blue-green in color. *V. vulnificus*, *V. mimicus*, and *V. harveyi* also look like this. Other *Vibrio* spp., including *V. cholerae* and an occasional *V. vulnificus*, have larger colonies and are yellow due to sucrose fermentation.
2. If no suspicious colonies are observed, the MPN is all negative (0/3, 0/3, 0/3, 0/3) and is recorded as <3 MPN/g based upon Table 4.1 in Laboratory 4.
3. It is important to record the plate and dilution of all enrichments that contain suspicious colonies. You do not need to count the colonies; however, you should record (in Table 8.1) the number of prospective positive plates from each dilution. For example: 10^{-1} 2/3, 10^{-2} 1/3, 10^{-3} 0/3, 10^{-4} 0/3 (refer to Laboratory 4).
4. Select two typical *V. parahaemolyticus* colonies from each prospective positive TCBS plate. Assign a code to each colony selected that will refer to the original dilution and test tube number. For example, isolate A from tube 1, or the 10^{-1} dilution could be labeled A11, while isolate two from the same tube and dilution could be B11.
5. For each colony, you need to patch streak onto GA and GSA, and streak for isolation on TSA + NaCl. Divide plates of GA and GSA into four quadrants using a permanent marking pen. Number the quadrants to correlate with your suspicious colonies. Also, label four plates of TSA + NaCl.
6. Perform a patch streak. Using a flamed, cooled loop, touch the suspicious colony, then make a heavy streak within the appropriate quadrant on GSA, then use the same inoculum on the appropriate quadrant on GA. Use the same inoculum for the first quadrant of a streak for isolation on TSA + NaCl.
7. Incubate plates at 35 to 37°C for 18 to 24 h.

Class 4

1. Read the GSA and GA plates. Halophilic *Vibrio* spp. will grow on the GSA plates but not on the GA plates. Because most are gelatinase positive, look for an opaque halo around the patch of growth. Record the results in Table 8.2.
2. Observe the TSA + NaCl plate. Are all the colonies similar? It is important to have a pure culture for biochemical testing. *Vibrio* spp. cultures can have two colony morphologies that may be stable or unstable. If more than one morphology are observed, each colony morphology should be streaked again to confirm purity.

3. Select colonies for further testing, and perform a Gram stain. *V. parahaemolyticus* is a Gram-negative rod-shaped organism. Record the results in Table 8.2.

4. Use a portion of the GSA patch to perform the oxidase test, which is performed as follows: Moisten a section of filter paper with liquid oxidase reagent (1% tetramethylparaphenylenediamine dihydrochloride). This should be performed in a chemical fume hood. Pick a small amount of fresh growth from the agar, and rub it onto the filter paper with a sterile toothpick. (Nonplatinum loops can interfere with this test.) Observe for a color change at exactly 10 sec. Dark purple is positive; no color change is negative. Record the results in Table 8.2.

5. If no isolates had the phenotypes of Gram-negative rod, growth on GSA but not GA, and oxidase positive, all tests are negative and testing should end here. If any Gram-negative rod-shaped isolates displayed growth on GSA but not GA, and where they are oxidase positive, the following media should be inoculated:

   a. TSI agar + NaCl: Touch a colony with a sterile needle and streak slant then stab butt of agar. Incubate at 18 to 24 h at 35 to 37°C. Tubes should be read within this time frame, because the results can change over time.

   b. MTM: Touch the same colony with a sterile needle, and stab the center of MTM to a depth about 2/3 from the top of the medium. Incubate 18 to 24 h at 35 to 37°C.

   c. Test for O/129 vibriostat sensitivity. Select the remaining portion of the colony and suspend well in 0.5 ml 2% NaCl. Wet a sterile swab in the bacterial suspension, squeeze out excess liquid on the side of the test tube, then streak over the surface of a TSA + NaCl plate. This will create a uniform lawn of bacterial growth. After the liquid is absorbed, add O/129 vibriostat sensitivity disks containing 150 µg and 10 µg (Oxoid cat. No. DD14 and DD15, respectively). Incubate at 18 to 24 h at 35 to 37°C.

Class 5

1. Read the results for the MTM media. Motility-positive cultures have a diffuse circular growth radiating from the inoculum stab. *V. parahaemolyticus* is motile. Record the results in Table 8.2.

2. Observe the results for TSI. *V. parahaemolyticus* exhibits a red alkaline slant (K) and an acid butt (A) without gas or H_2S production. Record the results in Table 8.2.

3. Read the results for O/129 vibriostat sensitivity. Any zone of nongrowth (clearing) around a filter paper disk is considered sensitive. *V. parahaemolyticus* is resistant to 10 µg but is sensitive to 150 µg.

4. At this point, study the results in Table 8.2. If any of the four colonies selected from TCBS agar for each MPN dilution had the typical reactions for *V. parahaemolyticus* (Table 8.3), the original MPN enrichment tube should now be labeled "presumptive *V. parahaemolyticus*." Use the data to calculate the MPN, then record it in Table 8.1. Use Table 4.1 in Laboratory 4 to calculate your result.

Further biochemical confirmatory tests include the O-Nitrophenyl-β-D-galactoside (ONPG) test, salt tolerance, growth at 42°C, Voges–Proskauer test, carbohydrate fermentation, urea hydrolysis, and testing for the production of thermostable direct hemolysin (Kanagawa phenomenon).

TABLE 8.1

Results of MPN

	Dilution				
	10^{-1}	10^{-2}	10^{-3}	10^{-4}	**MPN/g**[a]
Number of TCBS plates with suspicious colonies					NA
Number of presumptive tubes for MPN (calculated from data in Table 8.2)					

[a] Use data from Table 4.1 in Laboratory 4 for MPN/g estimate.

■ IV. RESULTS

T A B L E 8 . 2
Results of Isolate Testing

MPN Dilution	Tube Number	Isolate Code	Growth on GA	Growth on GSA	Oxidase Test	Gram Reaction	TSI Reaction	Motility	O/129 Sensitivity 10 µg	150 µg
10^{-1}	1									
	2									
	3									
10^{-2}	1									
	2									
	3									
10^{-3}	1									
	2									
	3									
10^{-4}	1									
	2									
	3									

TABLE 8.3

Biochemical Reactions to Differentiate Select Members of the *Vibrio* Genus[15]

Test	*V. parahaemolyticus*	*V. vulnificus*	*V. cholerae*
Gram stain	Gram-negative rods	Gram-negative rods	Gram-negative rods
Motility[a]	+	+	+
Oxidase[b]	+	+	+
Phenotype on TCBS	Green/blue-green	Green/blue-green	Yellow
GA	No growth	No growth	Growth
GSA	Growth	Growth	Growth
TSI[c]	K/A/–/–	K/A/–/–	A/A/–/–
O/129 sensitivity[d]			
10 µg	R	S	R
150µg	S	S	S

[a] Growth radiating from center stab is regarded as positive.

[b] A dark purple change when growth is spotted on oxidase reagent saturated filter paper is positive; no color change is negative.

[c] Slant/butt/gas/H_2S.

[d] Any zone of nongrowth (clearing) around a filter paper disk is considered sensitive.

Source: Elliot, E.L., Kaysner, C.A., Jackson, L., and Tamplin, M.L., in Bacteriological Analytical Manual, 8th ed., AOAC, Gaithersburg, MD, 1995.

■ V. DISCUSSION QUESTIONS

1. Why would you want to quantify the level of this human pathogen in a food product?

2. What are the advantages to performing an enrichment MPN compared to performing a selective plate count for quantification?

3. Is this an accurate method of quantification?

4. Speculate why this organism requires NaCl for growth.

LABORATORY NOTES

◼◼ LABORATORY NOTES

LABORATORY NOTES

LABORATORY 9

ISOLATION OF *CAMPYLOBACTER JEJUNI/COLI*

I. OBJECTIVE

- Learn the challenges involved in the basic enrichment and isolation of *Campylobacter jejuni/coli* from food.

II. BACKGROUND

Campylobacter jejuni and *Campylobacter coli* are Gram-negative, curved or spiral motile rods with a single polar flagellum that requires an atmosphere with reduced O_2 and increased CO_2. The optimal atmosphere for *Campylobacter* growth is 5% O_2, 10% CO_2, and 85% N_2; thus, this organism is considered a capriophile (carbon dioxide loving). In foods, this organism is not likely to grow under typical storage conditions.[16] The optimum growth temperature is 42 to 43°C, and its growth range is between 30 to 44°C.[16] These organisms are fairly frail in food and do not survive well outside their host's intestinal tract. These organisms are sensitive to freezing and die at room temperature. However, survival is increased at refrigeration temperatures, and food for analysis should be held at 4°C (and not be frozen). Even greater survival of *Campylobacter* organisms was found when food samples were stored under 100% N_2 with 0.01% sodium bisulfite at 4°C.[16] Analyses of food samples should be performed as soon as possible to prevent overgrowth of *Campylobacter* by psychrotrophic microflora.[17]

C. jejuni is almost exclusively associated with warm-blooded animals. A large percentage of all major meat animals have been shown to have this organism in their feces.[18] In general, fresh poultry tends to have a higher carriage of this organism than other meats.[18]

There are multiple methods that can be used to grow *Campylobacter* under a microaerobic atmosphere. The ideal method is to evacuate air and then fill a growth chamber or inflate a sterile bag from a gas canister with the appropriate gas mixture (5% O_2, 10% CO_2, and 85% N_2). Alternative methods include using commercially available gas generation packs for *Campylobacter* (such as those manufactured by BBL and Oxoid) or using low levels of anaerobic gas generation packs.[16] Both systems utilize single-use gas generator envelopes and catalyst pellets (palladium). To use one of these systems, the samples are placed in the "anaerobe" jar system. Depending upon the size of the jar (and type of generator envelope), one to three generator envelopes may be needed to create the microaerobic environment. The generator envelopes are cut open and placed in the jar in a vertical position. Then, 10 ml of water are added to each, and the jar is sealed. Once the jar is sealed, the water in the envelope reacts with a sodium borohydride tablet, producing hydrogen that subsequently combines with oxygen in the presence of the palladium catalyst producing water (thus reducing O_2 in the atmosphere). The generator envelope also has a sodium bicarbonate and citric acid tablet, which in the presence of water produces CO_2. The entire jar system is placed into an incubator.

Others proposed a less expensive alternative to generate a reduced O_2 and an enriched CO_2 environment within a bag. This alternative utilizes a combination of Alka-Seltzer® in water to generate CO_2 and steel wool soaked in copper sulfate to reduce gaseous oxygen by catalyzing an oxidation reaction of iron in the steel wool.[19] Although not commonly used in food laboratories, this method may be an economical alternative for classrooms.

Bacteriological Media Used in this Lab[17]

Ferrous sulfate, sodium bisulfate, and pyruvate supplement (FBP): This is a media supplement that increases the aerotolerance of *C. jejuni/coli*.[16] Each component is added to media so that there is a final concentration of 0.025%.[16] The stock solution is filter sterilized and then added to tempered media (47°C) after autoclaving. It is light sensitive, so stock solutions and media prepared with FBP should be stored in the dark.

Hunt enrichment broth (HEB): This is a selective media. Nutrients are supplied by a nutrient broth #2 (Oxoid) and yeast extract. Lysed horse blood (a hemin source) and FBP are added to the basal media after autoclaving to reduce the effects of oxygen in the medium. Selective antibiotics are also added to tempered autoclaved media and include vancomycin (to inhibit growth of aerobic and anaerobic Gram-positive organisms), trimethoprim (to inhibit many Gram-positive cocci and Gram-negative organisms), cefoperazone (to inhibit growth of *Enterobacteriaceae*, and *Pseudomonas*), and cycloheximide (to inhibit fungal growth).

Modified *Campylobacter* charcoal differential agar (mCCDA): This media uses commercially available CCDA (Oxoid CN739) with the addition of 2% yeast extract and CCDA selective supplement (SR155), which contains sodium cefoperazone (to inhibit growth of *Enterobacteriaceae* and *Pseudomonas*), rifampicin (to inhibit Gram-positive cocci), and amphotericin B (to inhibit fungal growth). In addition, charcoal and FBP are added as supplements to increase the oxygen tolerance of *C. jejuni*.

Brucella–FBP (BFBP) broth and agar: These are nonselective media used for obtaining general growth, checking purity, and evaluating sensitivity to antibiotics. FBP is added after autoclaving to tempered prepared brucella broth or agar (a rich media for fastidious organisms).

Oxidation-fermentation (O-F) medium for *Campylobacter*[16]: This is a semisolid nutrient (caitone) media that utilizes phenol red as a pH indicator. However, this media will be yellow in color when removed from the microaerobic atmosphere due to CO_2 absorption.[16] Therefore, after removing tubes from the microaerobic atmosphere, tubes should be kept at room temperature 1 to 2 h prior to reading to allow CO_2 diffusion. Glucose fermentative organisms will produce acid, and the medium will change to a yellow color (positive). The color of the medium will not change with the growth of a nonfermentative organism. *Campylobacters* are nonfermentative.

■ III. METHODS

NOTE: **In this laboratory, you may isolate a serious human pathogen. Therefore, each step should be treated as if that pathogen is present. As a precaution, you should wear eye protection and gloves during all steps after the initial washing of the chicken carcass.**

Class 1

Sample Preparation and Enrichment

Rather than homogenize solid foods in enrichment broth for *Campylobacter* isolation, foods are often gently rinsed in enrichment broth and particulates are removed by filtration. For whole carcasses, the bacteria in the rinse water are concentrated by centrifugation (Figure 9.1[17]).

Procedure

1. Place raw (never frozen) whole chicken carcass into a large sterile plastic bag, and add 200 ml 0.1% peptone water. Twist the bag to seal it, and shake the contents for 2 min.

2. Sanitize one corner of the bag with 70% ethanol, and then rinse with sterile distilled water. Cut aseptically using a sterile knife or scissors. Pour the rinse through a sterile cheesecloth-lined funnel into a sterile 250 ml (or larger) centrifuge bottle.

3. Centrifuge the wash solution for 15 min at 16,000 × g. Carefully remove the liquid, and suspend the pellet in 10 ml 0.1% peptone water.

4. Place 1 ml of rinse concentrate into 100 ml HEB in a sterile bag. Also, inoculate a control pure *C. jejuni* culture into HEB.

5. Ideally, bags should be inflated with Campy gas mixture (5% O_2, 10% CO_2 and 85% N_2) through a sterile filter.[4,17] Alternatively, if an appropriate gas mix is not available, loosely seal the bag with a twist tie, place the bottom in a beaker (for stabilization), and place this in an anaerobic chamber system. BBL™ GasPak™ 150 systems should use three CampyPak™ envelopes (5 to 12% CO_2, 5 to 15% O_2) or one GasPak anaerobe envelope (final CO_2 concentration approximately 5%).[4]

6. Incubate the enrichment at 37°C with shaking (100 rpm) for 4 h. If using an anaerobic chamber, place the entire chamber on the shaker.

7. After 4 h, transfer to a 42°C incubator and continue incubation with shaking (100 rpm) for 20 h.

Class 2

Streak Enrichment and Control Cultures onto Selective Differential Media

Procedure

1. Prepare dilution of enrichment and control cultures by swirling a swab in the enrichment broth, then pressing it against the side of the vessel to remove excess liquid. Break off the tip into a test tube with 9.9 ml 0.1% peptone water and vortex gently.

2. Streak undiluted and diluted enrichment and straight and diluted control cultures onto mCCDA.[17] This media should be dried prior to use. Inoculate plates by wetting a swab (as described above) and then swabbing approximately 40% of an mCCDA plate. Use a loop to streak for isolation from the swabbed area.

3. To prevent swarming of colonies, add a sheet of filter paper wetted with two to three drops of glycerol (to absorb excess liquid) to the anaerobe jar before sealing.

4. Incubate at 42°C for 24 h under microaerophilic conditions in a BBL GasPak 150 system. If Campy gas mixture (5% O_2, 10% CO_2, and 85% N_2) is available, evacuate jars and fill with the mixture. If not, use Campy envelopes as described in step 7. If no growth is observed after 24 h, continue incubation under the same conditions for an additional 24 to 48 h.

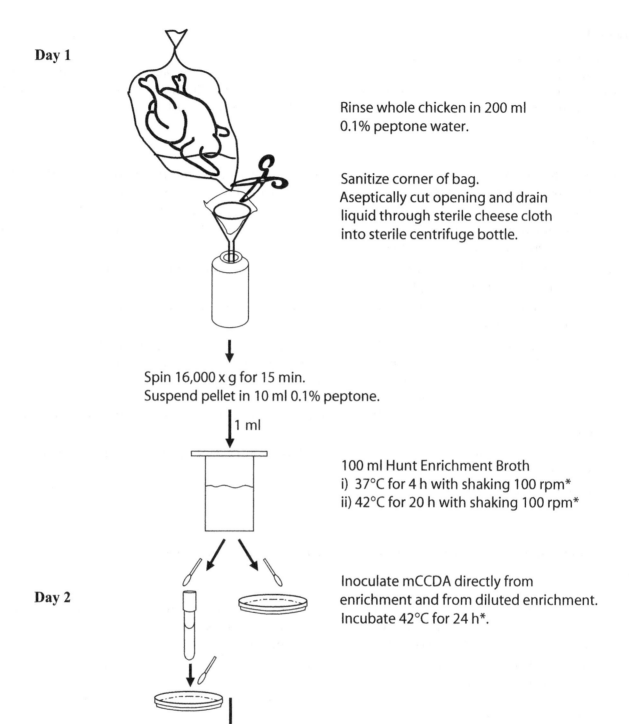

Day 1

Rinse whole chicken in 200 ml 0.1% peptone water.

Sanitize corner of bag. Aseptically cut opening and drain liquid through sterile cheese cloth into sterile centrifuge bottle.

Spin 16,000 x g for 15 min. Suspend pellet in 10 ml 0.1% peptone.

1 ml

100 ml Hunt Enrichment Broth
i) 37°C for 4 h with shaking 100 rpm*
ii) 42°C for 20 h with shaking 100 rpm*

Day 2

Inoculate mCCDA directly from enrichment and from diluted enrichment. Incubate 42°C for 24 h*.

Days 3, 4 and 5   Confirm cellular morphology of typical *C. jejuni/coli* colonies. Inoculate media for confirmation tests.

*ALL Incubation steps must be performed under microaerobic conditions: 5% O_2, 10% CO_2, 85% N_2

Figure 9.1   Flowchart of steps used for isolation of *Campylobacter jejuni/coli* from a whole chicken.

Class 3

1. Observe the colonies on mCCDA. Colonies of *Campylobacter* can have the following appearance: "smooth, shiny, and convex with a defined edge, or flat, transparent or translucent, and spreading with an irregular edge; colorless to grayish or light cream; and usually 1 to 2 mm in diameter but may be pinpoint to several mm in diameter."[17]

2. Pick a small portion of a colony and suspend in a drop of water or saline on a glass microscope slide to confirm cellular morphology prior to inoculating media for confirmational tests. Use a phase contrast microscope with a 100× oil immersion objective to visualize cellular morphology. Young cells of *Campylobacter* are narrow curved rods (0.2 to 0.8 μm wide by 1.5 to 5 μm long) with a darting or corkscrew-like motility. Sometimes, pairs can look like an "S" or "gull wings." Cells grown for more than 72 h may appear as coccoid (and may be nonculturable).

3. Colonies with typical *Campylobacter* cellular morphology should be used for confirmatory tests. Pick up to four suspect colonies (one of each colony morphology) into BFBP broth. Also run control culture. Incubate with caps loosened for 24 to 48 h at 42°C with an appropriate Campy gas mixture.

Class 4

Confirmation Tests[17]

1. Test for glucose fermentation by inoculating a tube (O-F) medium[16] from BFBP broth. Incubate under microaerobic conditions at 42°C for 1 to 3 days.

2. Test for susceptibility to nalidixic acid/cephalothin by inoculating a BFBP agar plate with 0.2 ml of the BFBP broth growth, and spread with a glass rod. Drop disks of nalidixic acid (30 μg) and cephalothin (30 μg) on a spread plate and press with a sterile forceps to assure adherence to the agar surface. Incubate under microaerobic conditions at 42°C for 1 to 3 days.

Class 5

Record your results in the results section:

1. Glucose fermentation test: O-F media will be yellow in color when removed from the microaerobic atmosphere due to absorption of CO_2.[16] Therefore, after removing test tubes from the microaerobic atmosphere, tubes should be kept at room temperature 1 to 2 h prior to reading. Glucose fermentative organisms will produce acid, and the medium will change to a yellow color (positive). The color of the medium will not change with the growth of a nonfermentative organism. *Campylobacters* are nonfermentative.

2. Susceptibility to nalidixic acid and cephalothin: Observe the lawn of growth around each antibiotic disk. A clear zone of any size indicates sensitivity to the antibiotic. The lawn of growth may be very light, so adjusting the angle of the plate in relation to the light may be useful. *C. jejuni* and *C. coli* are both resistant to cephalothin and sensitive to nalidixic acid. This plate will also be used for the oxidase and catalase tests.

3. Use a portion of the BFBP agar plate (to perform the oxidase test): Moisten a section of filter paper with liquid oxidase reagent (1% tetramethylparaphenylenediamine dihydrochloride). This should be performed in a chemical fume hood. Heavily smear growth from agar and rub it onto the filter paper with a sterile toothpick or platinum or plastic loop. (Nonplatinum loops can interfere with this test.) Observe this for a color change within 30 sec. Dark purple is positive; no color change is negative. Members of *Campylobacter* spp. are oxidase positive. Record your results in the results section.

4. Catalase test: After performing all other tests, add two to three drops of 3% H_2O_2 to the surface of the BFBP plate. Examine after 1 to 10 min for formation of bubbles due to generation of O_2 in the presence of the enzyme catalase. *C. jejuni* and *C. coli* are both catalase positive.

5. All isolates that are nonfermentative, catalase positive, resistant to cephalothin, sensitive to nalidixic acid, and oxidase positive are considered to be *C. jejuni/coli*.

■ IV. RESULTS

Results Worksheet (Record Results of this Laboratory)

Test	Food Isolates	Pure Culture Control
Colony morphology on mCCMA	A	
	B	
	C	
	D	
Cellular morphology	A	
	B	
	C	
	D	
Nalidixic acid sensitivity	A	
	B	
	C	
	D	
Cephalothin sensitivity	A	
	B	
	C	
	D	
Oxidase test	A	
	B	
	C	
	D	
Catalase test	A	
	B	
	C	
	D	

■ V. DISCUSSION QUESTIONS

1. What are the different methods of modifying the atmosphere to allow for *Campylobacter* growth?

2. Because the atmosphere is critical for growth of this organism, how can you measure your atmosphere for "quality control" purposes? In other words, if you get a negative result, how do you know it is due to the absence of *Campylobacter* spp. cells rather than a failure of having the appropriate atmosphere?

3. Discuss any difficulties you experienced working with pure cultures of *Campylobacter*. Discuss how these difficulties are compounded in recovering *C. jejuni/coli* from foods.

LABORATORY NOTES

LABORATORY NOTES

LABORATORY 10

ENUMERATION OF *STAPHYLOCOCCUS AUREUS* FROM FOOD

I. OBJECTIVE

- Become familiar with the isolation and enumeration of *Staphylococcus aureus* from foods.

II. BACKGROUND

S. aureus is present on the skin and in the nasopharnyx area of humans and animals. Often, *S. aureus* will get into food from food handlers, from animal skin, or from dirty food preparation surfaces. *S. aureus* cannot grow at refrigeration temperatures and is a relatively poor competitor with other food microflora. *S. aureus* growth usually occurs during temperature abuse. Temperature abuse occurs when food is kept in the temperature danger zone — 4.4 to 60°C (40 to 140°F) — for prolonged periods of time. The optimal temperature for *S. aureus* growth is between 18 to 40°C (64 to 104°F). Because *S. aureus* grow poorly in the presence of other food microflora, these organisms tend to grow better in cooked or processed foods. In addition, *S. aureus* can grow at lower a_w compared to most other bacteria (down to a_w 0.85). Therefore, *S. aureus* may grow under a reduced a_w or on high salt foods that will inhibit the growth of most pathogens as long as the temperature permits growth. Metabolically, *S. aureus* can utilize mannitol, which is not seen with other staphylococcal species, such as *Staphylococcus epidermidis*. The production of the enzyme coagulase by a majority of *S. aureus* strains can also be used to differentiate this species from other staphylococci.

Some strains of *S. aureus* have the ability to produce staphylococcal enterotoxins (SEs) while growing in foods. The main symptoms of SE intoxication are vomiting and diarrhea (without fever) 4 to 12 h after consumption. Foods commonly associated with staphylococcal food poisoning are deli meats (especially ham), deli salads (ham, chicken, potato), and cream puffs. Note that these foods all undergo preparation steps involving human handling after processing. In addition, these products may potentially undergo temperature abuse, either before or after sale to the consumer.

Large numbers of enterotoxigenic *S. aureus* (greater than 10^6 CFU/g) are needed to produce enough staphylococcal enterotoxin to cause this type of food intoxication. Although it is desirable to have no *S. aureus* in food products, small numbers of *S. aureus* (less than 10^3 CFU/g) are often present. Because large numbers are required to produce detectable quantities of SE, *S. aureus* is often enumerated in food products as an indication of temperature abuse and the potential for *S. aureus* to cause foodborne disease.[18]

It is important to remember that SE is more stable and heat resistant than *S. aureus* cells. Therefore, low levels of *S. aureus* do not always assure the product has SE. Possibly, there were high levels of *S. aureus* growth at one time, and subsequent processing eliminated viable *S. aureus* but did not eliminate the biological activity of the enterotoxins. For this reason, direct detection of SE using immunological tests or detection of other metabolites (staphylococcal heat-stable thermonuclease) can be used as an indication that high levels of *S. aureus* were present in the food product.[20]

Media Used in this Lab

Baird–Parker medium: This is the preferred selective/differential medium for *S. aureus* detection.[20] This medium contains tryptone and beef extract as basic nutrients, along with glycine and sodium pyruvate to encourage growth. The selective agents are lithium chloride and potassium tellurite, which will inhibit most Gram-negative organisms and slow to inhibit the growth of most other Gram-positive organisms. Egg yolk is added for the detection of the enzyme lecithinase produced by a large number of *S. aureus* strains. Typical coagulase-positive staphylococci have the ability to reduce potassium tellurite-producing black colonies and produce the enzyme lecithinase, which causes a visible clearing around the colonies.

Mannitol salt agar: This is an alternative selective media for staphylococci. This medium contains peptone, beef extract, mannitol, NaCl, agar, and phenol red. The high salt concentration (7.5%) inhibits the growth of most other organisms, and the only fermentable carbohydrate is mannitol. On this medium, *S. aureus* appears as a yellow colony, while mannitol-negative nonpathogenic *S. epidermidis* will appear as a red colony.

Brain–heart infusion broth: This medium is made of infusions from calf brains, beef heart, peptone, glucose, and NaCl. This medium is highly nutritive and is used to produce heavy growth of fastidious organisms.

Coagulase plasma with ethylenediaminetetraacetic acid (EDTA): This is purchased in a lyophilized form. EDTA is added as an anticoagulant. The reconstituted plasma is used to determine if a staphylococcal species has the ability to produce the enzyme coagulase, which will cause clotting of rabbit plasma.

■ III. METHODS

Class 1

Sample Preparation

Procedure

1. Use a sterile spatula, weigh 10 g of ham salad, and place in a sterile Stomacher bag.
2. Add 90 ml of sterile saline. This is your first 1/10 dilution.
3. Homogenize in a paddle blender (such as a Stomacher) for 2 min.
4. Prepare duplicate spread plates on mannitol salt agar or Baird–Parker medium using the dilution scheme shown in Figure 10.1.
5. Incubate plates for 24 h at 37°C.

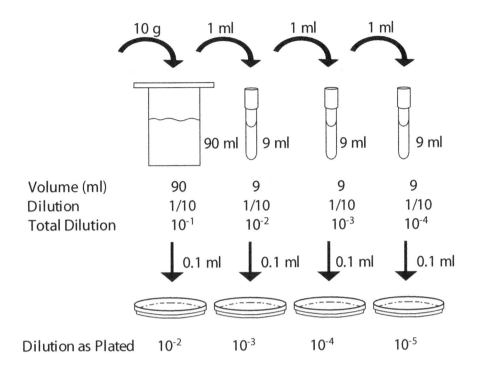

Figure 10.1 Dilution scheme suggested for this laboratory.

Class 2

Results

1. Observe plates. Count yellow colonies on mannitol salt agar or black colonies on Baird Parker medium, and calculate the presumptive "*S. aureus* CFU/g."

2. For confirmation, pick two representative presumptive *S. aureus* colonies. Use a flamed loop and transfer each colony into 2 ml brain–heart infusion broth (1 colony/tube of broth). Incubate at 37°C for 24 to 28 h.

Class 3

Coagulase Test for Confirmation of S. aureus

Procedure

1. Add 0.2 ml of 24 to 28 h culture to 0.5 ml of coagulase plasma with EDTA in 10 × 75 mm tubes.

2. Incubate in a 37°C water bath, and examine periodically for clot formation. *S. aureus* should clot the plasma within 4 h.

3. When the representative colonies are coagulase positive, the count becomes confirmed for *S. aureus*.

Further characterization tests could include anaerobic utilization of glucose and mannitol (*S. aureus* are usually positive for both of these tests), sensitivity to lysostaphin, and thermonuclease production.[20]

■ IV. RESULTS

Dilution as Plated	CFU/Plate	CFU/Plate	Average CFU/Plate
10^{-2}			
10^{-3}			
10^{-4}			
10^{-5}			

CFU/g =

Record the results of the coagulase test:

■ V. DISCUSSION QUESTIONS

1. Why is direct selective or differential plating used for *S. aureus* testing rather than selective enrichment isolation?

2. There is a report of a foodborne outbreak due to consumption of cheddar cheese. The victims have the classic symptoms of staphylococcal food poisoning. The symptoms include vomiting and diarrhea without fever within 4 h after consumption of the cheese. The symptoms only last a short time (8 to 12 h). As the good food microbiologist, you plate the product and find that there are no detectable *S. aureus*. How could you confirm or eliminate SEs as the cause of this outbreak?

3. How could you confirm the presence or absence of SE in a food product?

LABORATORY NOTES

▆ LABORATORY NOTES

LABORATORY 11

ISOLATION OF *LISTERIA* SPP. FROM REFRIGERATED FOODS

I. OBJECTIVE

- Use current culture techniques to isolate *Listeria* spp. from foods.

II. BACKGROUND

Listeria monocytogenes is a foodborne pathogen. It can easily be isolated from many sources in the environment as well as from a wide variety of fish, birds, and mammals. Because *L. monocytogenes* is a psychrotroph, the growth of this organism cannot be controlled by refrigeration temperature alone.

This organism is thought to cause a mild flu-like illness in healthy adults, but it can cause serious illness in young children, the elderly, and the immunocompromised. Patients with serious listeriosis can develop septicemia, meningitis, and endocarditis.[21] This is a serious disease, with a mortality rate of around 20%, if full listeriosis develops.[21] Pregnant women can have mild flu-like symptoms, while their fetuses can develop serious infections *in utero*, causing abortion, stillbirth, or an infected newborn with listerial meningitis or septicemia.

NOTE: **We may or may not isolate this organism from foods. However, if you have an underlying health condition, are pregnant, or are uncomfortable with this exercise, notify your instructor prior to class, and an alternate assignment will be arranged.**

Due to the seriousness of listeriosis to susceptible populations, the U.S. Food and Drug Administration (FDA) and the U.S. Department of Agriculture (USDA) set a "zero tolerance" of *L. monocytogenes* for many food items. When "detection" of the organisms is more important than "enumeration," isolation procedures involve selective enrichment steps. Selective enrichment usually involves adding an amount of sample to a selective enrichment broth. The selective enrichment broth allows growth of the organism of interest while it inhibits competitive organisms. After selective enrichment, selective or differential media are used to isolate the organisms. Confirmation is performed by various biochemical tests. There are two isolation methods for listerial detection: the FDA enrichment method used to isolate *L. monocytogenes* from dairy and seafood products,[22] and the USDA enrichment method to isolate *L. monocytogenes* from meat products.[23] Because *Listeria* spp. can grow at refrigeration temperatures, we will test a number of different refrigerated foods using the USDA enrichment method. Each group will test 25 g of a product. When government labs test products, they enrich 13 separate 25 g samples for a total of 325 g of tested product.

Bacteriological Media Used in this Lab[22]

University of Vermont broth (UVM 1° selective enrichment): This selective enrichment contains tryptose, beef extract, yeast extract, NaCl, sodium phosphate, potassium phosphate, esculin, nalidixic acid, and acriflavin. Phosphate is a buffering agent. Nalidixic acid, lithium chloride, and acriflavin are added at a level that will not inhibit *Listeria* spp. but will inhibit other background microflora.

Fraser broth (2° selective enrichment): This broth medium contains tryptose, beef extract, NaCl, phosphate buffer, lithium chloride, esculin, nalidixic acid, and acriflavin. Nalidixic acid, lithium chloride, and acriflavin are added at a level that will not inhibit *Listeria* spp. but will inhibit other background microflora. *Listeria* spp. have the ability to utilize esculin as the only carbohydrate source. When esculin is hydrolyzed, the product reacts with ferric compounds to produce a black color.

Oxford agar (selective or differential colony identification): Oxford agar is made from Columbia blood agar base (a rich medium), esculin, ferric ammonium citrate, and lithium chloride. This medium is supplemented with antibiotics and antimicrobial compounds (cycloheximide, colistin sulfate, acriflavin, cefotetan, and phosphomycin) to inhibit the growth of most organisms other than *Listeria* spp. Again, *Listeria* spp. can hydrolyze the esculin in the media, and the hydrolysis product can react with the ferric ammonium citrate in the media, causing listerial colonies to appear shiny black surrounded by black halos. Enterococci that can hydrolyze esculin and blacken the agar may be also present. However, because they grow slowly on this medium (a weak esculin reaction is obvious after 40 h of incubation), as the incubation period is kept under 30 h, the growth of enterococci should not be a problem.

Confirmation of *Listeria* spp.

Observation by Henry illumination: This test is commonly used to identify listerial colonies. Light is shined through plates at a 45° angle. *Listeria* colonies appear blue-gray to blue. Nonlisterial colonies tend to appear as yellowish or orange.

Tumbling motility: Colonies suspected of being *Listeria* spp. are examined using a wet mount and a phase contrast microscope. Typical *Listeria* spp. have a gentle rotating tumbling motion. Cocci, large rods, and rods with rapid swimming motions are rejected. Generally, the motility is expressed at lower temperatures, so incubation is at 30°C or below.

■ III. METHODS

NOTE: **In this laboratory, you may isolate a serious human pathogen. As a precaution, you should wear eye protection and gloves during all steps after the initial homogenization of the product in enrichment broth.**

Class 1

Sample Preparation and Primary Enrichment

1. Aseptically open bags of refrigerated foods by sanitizing the outsides of bags with 70% ethanol. Cut open the outside of the bag with a sterile knife.
2. Label the Stomacher bag with your name and food product.
3. Place a sterile paddle blender (Stomacher) bag on a balance and tare weight. Aseptically cut and add 25 g of product pieces to the bag on the balance.
4. Add 225 ml of UVM broth, and blend for 2 min.
5. Loosely close the bag, and place it in a beaker. Incubate the primary enrichment at 30°C for 24 h.

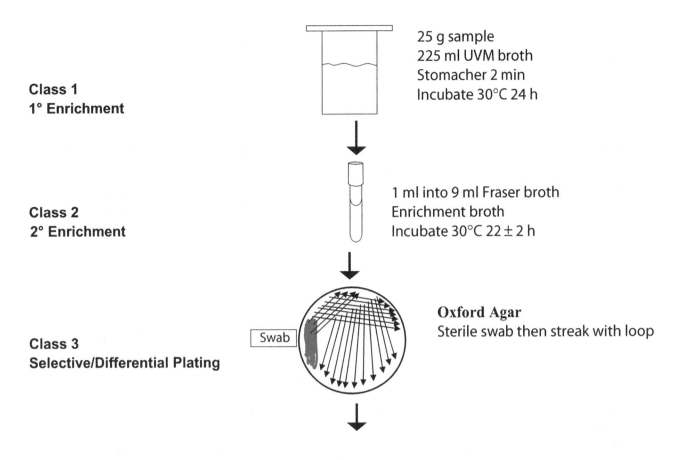

Class 1
1° Enrichment

25 g sample
225 ml UVM broth
Stomacher 2 min
Incubate 30°C 24 h

Class 2
2° Enrichment

1 ml into 9 ml Fraser broth
Enrichment broth
Incubate 30°C 22 ± 2 h

Class 3
Selective/Differential Plating

Swab

Oxford Agar
Sterile swab then streak with loop

Class 4 and 5
Presumptive
***Listeria* spp.**

Streak black colonies on TSAYE. Use for:
Observation with Henry Illumination
Gram Stain
Tumbling Motility
Catalase Test

Figure 11.1 Flowchart for *Listeria monocytogenes* isolation using the FSIS/USDA method. (From USDA/FSIS, *USDA/FSIS Microbiology Laboratory Guidebook*, 3rd ed. (http://www.fsis.usda.gov/ophs/microlab/mlgbook.htm), 1998.

Class 2

Secondary Enrichment

1. Transfer 0.1 ml of UVM broth to 10 ml of Fraser broth. Incubate at 35°C for 24 h. Incubate an inoculated tube of Fraser broth as a control.

Class 3

Selective and Differential Plating

1. Observe your Fraser enrichment. Blackening of the broth indicates growth of esculin-reducing organisms. Originally, the USDA isolation procedure was stopped if the Fraser broth was not blackened after 24 h. However, recent studies now suggest that Fraser broth that does not turn black after 24 h may still contain *Listeria* spp. Consequently, all tubes of Fraser broth (black or not) will be streaked onto Oxford agar.

2. Transfer the Fraser enrichment onto two plates of Oxford medium using a sterile swab. Make a thick inoculum with the swab in the first quadrant of the plate. Use a flamed loop to streak from the swabbed portion.

3. Incubate at 35°C for 24 h.

Class 4

1. Observe the Oxford agar plates. Listerial colonies on this media will appear as black colonies due to esculin reduction. If there are black colonies, there is presumptive *Listeria* spp.

2. If typical listerial colonies are observed, streak three colonies onto tryptic soy agar with 0.6% yeast extract (TSA–YE). These will be used for confirmatory tests. Incubate at 30°C for 24 h.

Class 5

Confirmatory Tests

1. Observe tumbling motility using a wet mount. Suspend a typical colony in saline on a microscope slide. Place a coverslip on top of the drop. The liquid should spread out under the coverslip. Observe this under the oil objective with a phase contrast microscope. *Listeria* spp. will appear as short rods that move with a tumbling motion.

2. Check the catalase reaction. Place a drop of 3% H_2O_2 on a colony. Listeria colonies are catalase positive.

3. Perform a Gram stain. *Listeria* are short Gram-positive rods. If cultures are older than 24 h, the Gram reaction can be variable.

4. Observe the colonies using Henry illumination. Look at the colonies with 45° transillumination to observe the representative blue color that is typical of *Listeria* spp.

If everything is positive at this point, the presence of *Listeria* spp. is confirmed.

In order to determine if the *Listeria* spp. belong to the species *L. monocytogenes* (human pathogen) or other nonpathogenic *Listeria* spp. (*L. seeligeri*, *L. ivanovii*, *L. innocua*, *L. grayi*, or *L. murrayi*), a number of other tests must be performed:

- *L. monocytogenes* are able to lyse sheep blood cells. This clearing is called β-hemolysis. *L. innocua* will not show a zone of hemolysis, *L. ivanovii* will usually give a well-defined clear zone, and *L. monocytogenes* and *L. seeligeri* show a slight clearing on just the edges of the colonies. Strain identification based on differences in hemolysis reaction must be further tested using the CAMP reaction.

- The Christie–Atkins–Munch–Peterson (CAMP) test looks at hemolysis in the presence of two other bacterial species (*Staphylococcus aureus* and *Rhodococcus equi*). In this test, *S. aureus* and *R. equi* are streaked in parallel on two edges of a sheep blood agar plate. *Listeria* spp. isolates are then streaked perpendicularly between the two strains. The plates are then incubated at 35°C for 24 to 48 h. The hemolysis of the test strains is examined for regions of enhanced hemolysis near *S. aureus* and *R. equi* growth. *L. monocytogenes* will show enhanced β-hemolysis in the presence of *S. aureus* but not in the presence of *R. equi* streak. *L. ivanovii* shows enhanced hemolysis from *R. equi* but not *S. aureus*.

- Metabolically, *L. monocytogenes* does not ferment mannitol or xylose but can ferment rhamnose.

- Polymerase chain reaction (PCR) may also be used to confirm *Listeria monocytogenes*, either by amplifying a specific region of the 16S rRNA (as performed in Laboratory 12) or by amplifying a portion of pathogenesis genes.

■ IV. RESULTS

Observed Darkening of Fraser Broth

Isolate Number	Colonial Phenotype on Oxford Agar	Gram Reaction	Catalase	Tumbling Motility	Henry Illumination	Presumptive *Listeria* spp.

■ V. DISCUSSION QUESTIONS

1. What is the theoretical limit of detection for this enrichment procedure?

2. Thirteen separate samples of the same product were tested. After testing, three of the 13 samples were confirmed for *L. monocytogenes*. What could you extrapolate about either the level of contamination or the homogeneity of the product? What would the FDA or USDA think about these results?

3. What are the advantages of using two selective enrichment procedures rather than just one?

■ LABORATORY NOTES

LABORATORY NOTES

■ LABORATORY NOTES

LABORATORY 12

SCREENING OF *LISTERIA* ENRICHMENTS USING PCR-BASED TESTING

■ I. OBJECTIVES

- Become familiar with using rapid methods to detect pathogens in foods.

- Obtain a basic understanding of polymerase chain reaction (PCR) for identifying bacteria in foods.

■ II. BACKGROUND

One of the greatest problems with testing for pathogens in food is the length of time it takes from enrichment to confirmation. In recent years, a large number of rapid methods (such as enzyme-linked immunosorbent assay [ELISA], PCR, gene probes, or lateral flow immunoassays) for identifying various pathogens have come onto the market. Typically, rapid methods are performed after the enrichment steps and prior to culture biochemical confirmation. They are used to rapidly assess a negative result. There is always a remote possibility that a nonviable cell could be bound by an immunoglobulin, or DNA from a nonviable cell could be amplified, giving a false-positive result. Therefore, when there is a positive result, it is prudent to continue with culture biochemical confirmation.

Despite the sensitivity of many rapid methods, most require an enrichment step prior to testing. As with a traditional enrichment, the main purpose of enrichment before using a rapid method is to increase the number of target cells in relation to background microflora. This is especially important because rapid methods are often performed in small volumes. For example, in this lab, you will use 5 µl of a secondary enrichment in 200 µl of cell lysis buffer. Only a portion (50 µl) of this lysate will be used for PCR amplification. Enrichments are also used before a rapid method to reduce the amount of interfering food components that carry over into the reaction and to help assure that viable cells outnumber nonviable cells (thus reducing false positives).

PCR is a method that produces many copies of a specific fragment of DNA. The food enrichment is used as the bacterial source from which the chromosomal DNA is extracted before performing PCR. This DNA extraction step is critical for the success of a PCR and can be performed by using a variety of methods (heating with or without proteases, using commercial extraction kits, etc.). Once the DNA is extracted, it is mixed with the following key components in a PCR cocktail:

- Forward and reverse DNA primers: These are short pieces of single-stranded DNA (15 to 20 bases in length) that have a specific sequence (homologous) that flank a target section of

DNA (Figure 12.1). *Taq* DNA polymerase will use these primers as a starting point to synthesize DNA in a $5' \rightarrow 3'$ direction.

- Nucleotides (dATP, dTTP, dCTP, and dGTP): These are the building blocks of DNA. *Taq* DNA polymerase will add the nucleotide homologous to the template at the $3'$ end of the primers to synthesize DNA in a $5' \rightarrow 3'$ direction.

- *Taq* DNA polymerase: This is the key component of PCR. In order for PCR to work efficiently, the DNA polymerase must be heat stable. The most common heat-stable DNA polymerase is *Taq* DNA polymerase, originally isolated from *Thermus aquaticus*, a thermophilic bacterium. This enzyme remains biologically active through multiple melting cycles (95°C) and has optimal activity at 72°C. There are other heat-stable DNA polymerases that can be used for PCR.

- Reaction buffer: This buffer keeps the PCR reaction at the optimal pH and magnesium chloride concentration for *Taq* DNA polymerase activity.

The PCR is performed in a thermal cycler, which is an automated heating block that can be programmed to rapidly change the incubation temperature. The PCR is controlled by changing the incubation temperatures and hold times. The three basic steps of PCR are melting, primer annealing, and elongation (Figure 12.1). During the melting step, the reaction is heated to 95°C for about 1 min. This causes the hydrogen bonds of the double-stranded DNA to separate (also called denaturation or "melting"). The next step is to anneal the primer to the target sequence. This is accomplished by dropping the temperature to, typically, 40 to 60°C for about 1 min. The annealing temperature is dependent upon the composition of the primers. Generally, using a higher annealing temperature will increase the specificity of the hybridization reaction, and the optimal annealing temperature can be determined experimentally. During the elongation step, the temperature of the reaction is increased to 72°C for the maximum rate of *Taq* DNA polymerase activity. This cycle of melting, annealing, and elongation is repeated for 30 to 35 cycles.

The first three cycles of PCR are illustrated in Figure 12.1.[24] The target sequence for the first cycle is the double-stranded DNA (dsDNA) added to the reaction by the researcher. However, during each subsequent cycle, the newly synthesized dsDNA from previous cycles becomes additional template DNA. In PCR, the expected amplified product length should be the distance between the two primers. It is important to note that Figure 12.1 is a schematic of early amplification cycles, showing that only two of the eight copies produced are the expected size after three completed cycles. After 30 or more cycles, using previously amplified fragment as template, the amplified DNA fragment will appear homogeneous on an agarose gel. The majority of the copies of the product will have a length equivalent to the distance between the two PCR primers.

PCR is a sensitive method. Theoretically, as little as one copy of the target sequence is enough to allow amplification of over 10^8 copies in 30 cycles. Because of this high sensitivity, it is important that your work area, pipetters, and thermal cycler do not become contaminated with DNA. To prevent DNA contamination, the work area where template DNA is prepared and manipulated should be separate from the electrophoresis area. Ideally, the cocktail should be prepared in a laminar flow or PCR hood sterilized with UV light. Before and after work, each work area, the equipment, and the thermal cycler block should be cleaned with 20% bleach solution and rinsed with water. Clean gloves and a lab coat should be worn at all times. Pipetters can be a source of contamination through aerosol from the pipetter, so a new sterile aerosol barrier tip must be used for every measurement. Never use a tip more than once. In order to assure that an amplification product is not contaminated, every PCR run should have at least one tube that is run without the addition of template DNA, as a negative control. If bands are visible from this reaction, it is an indication that DNA contamination occurred at some step in the process.

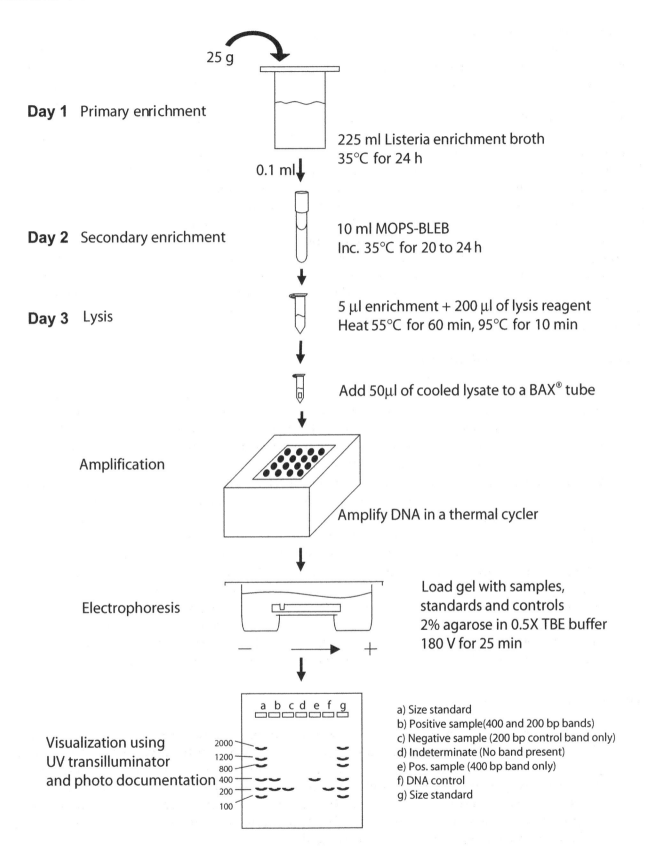

Figure 12.1 Identification of presumptive *L. monocytogenes*-positive foods using the BAX system.

In this lab, the food sample will undergo primary and secondary enrichment prior to testing with the BAX® system (Figure 12.2). The BAX system is a commercially available PCR-based test kit manufactured by DuPont Qualicon (Wilmington, DE). This kit includes lysis reagents, ingredients for the DNA cocktail (primers, DNA polymerase, and nucleotides) in tablet form in prepackaged PCR tubes, loading dye, and DNA ladder. After secondary enrichment, the enrichment sample undergoes lysis to release DNA from the bacterial cells. The lysate is then added to the BAX tablet, and amplification takes place in a thermal cycler with a two-temperature amplification program (denature at 94°C for 15 sec, anneal and elongate at 70°C for 3 min) optimized for this test kit. The BAX system for *Listeria monocytogenes* amplifies a proprietary target — a 400 base pair DNA fragment specific for this organism. In addition, there is a positive control incorporated into the tablet. This positive control is a band of DNA (200 bp) that should always be amplified (regardless of the presence of the target sequence), indicating that the PCR reaction was not inhibited (Figure 12.2). In the BAX system for *L. monocytogenes*, if the 400 bp band is present (with or without the 200 bp control band), the sample is considered to be presumed positive for *L. monocytogenes*. If the control band (200 bp) is present but the test band is lacking (400 bp), the sample is negative for *L. monocytogenes*. If no bands are visible, the sample is considered indeterminate and may indicate a problem with the PCR reaction (such as a malfunctioning thermal cycler or interference of the PCR reaction with food components).

Media/Chemicals Used in this Lab

Listeria enrichment broth (LEB; primary enrichment): This is the enrichment media recommended by the FDA.[4] This media contains trypticase soy broth and yeast extract as the nutrient base, phosphate buffering agents (pH 7.3), and selective agents. The selective agents are cycloheximide (to inhibit fungal growth), acriflavin (will inhibit growth of Gram-negative organisms), and nalidixic acid (to inhibit Gram-negative organisms).

MOPS-buffered listeria enrichment broth (MOPS-BLEB; secondary enrichment): MOPS-BLEB is the secondary enrichment recommended by DuPont Qualicon for use with the BAX system. This has the same composition as LEB, except the phosphate buffering agents are replaced with MOPS (3-morpholinopropansesulfonic acid), a synthetic buffering agent. With a pK_a of 7.2, MOPS is useful in buffering solutions with a pH range of 6.5 to 7.9.

Ethidium bromide: This is a fluorescent DNA dye that will be used in this lab to visualize DNA in agarose gels. It is an intercalating dye that binds to the phosphate backbone of nucleic acids. Extreme caution should be taken during its use. This dye is toxic and can be absorbed through the skin, mucus membranes, and eyes. Therefore, when handling gel and buffers containing ethidium bromide, you should wear suitable protective clothing (lab coat, gloves, and eye protection). In addition, this material and its container must be disposed of as hazardous waste.

■ III. METHODS

NOTE: **In this laboratory, you may isolate a serious human pathogen; therefore, each step should be treated as if that pathogen is present. As a precaution, you should wear eye protection, lab coats, and gloves during all steps. *L. monocytogenes* is especially dangerous to pregnant women and persons with underlying health concerns (such as those who are immunocompromised). If you have an underlying health condition, are pregnant, or are uncomfortable with this exercise, notify your instructor prior to class, and an alternate assignment will be arranged.**

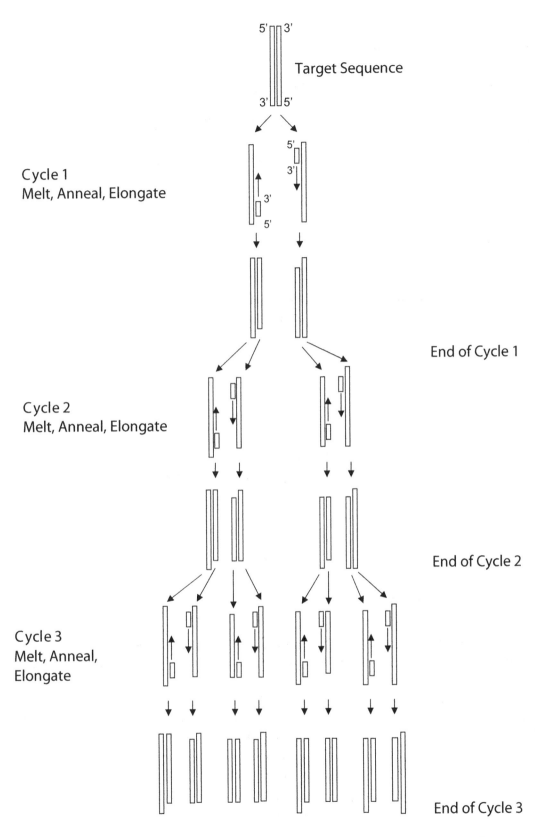

Figure 12.2   DNA amplification using PCR. (Adapted from Prescott, L.M., Harley, J.P., and Kein, D.A., *Micro-biology*, 2nd ed., Wm. C. Brown Publishers, Dubuque, IA, 1993.)

Class 1

Sample Preparation and Enrichment

Procedure

The methodology for this lab is outlined in Figure 12.2.

1. Perform primary enrichment according to food type. LEB is a general enrichment method recommended by the FDA.[22]
2. Measure 25 g of food product. Add this to 225 ml LEB in a Stomacher blender. Homogenize.
3. Incubate at 35°C for 18 to 24 h.

Class 2

Secondary Enrichment in MOPS-BLEB

Procedure

1. Take the primary enrichment and dilute 1:100 (0.1 ml into 10 ml) MOPS-BLEB.
2. Incubate one test tube of uninoculated MOPS-BLEB to use as a positive control.
3. Incubate at 35°C for 20 to 24 h.

Class 3

BAX Test

Procedure

1. Turn on heating blocks to 55°C and 95°C.
2. Program the thermal cycler (if needed) and perform during 55°C incubation. (Refer to the individual thermal cycler manual for programming specifications.)

   Program the thermal cycler for the BAX system:
   a. Initial step: Hold at 94°C for 2 min.
   b. Amplify at two temperatures for 38 cycles:
      i. Denature: 94°C for 15 sec.
      ii. Anneal/elongate: 70°C for 3 min.
   c. Final hold: 25°C, indefinitely.
3. Add protease to the lysis buffer (both supplied with kit). Add 12.5 µl of protease per 1000 µl of lysis buffer. You will need 200 µl lysis buffer per sample. One group should prepare enough lysis reagent for the entire class.
4. For DNA isolation, label one lysis tube for each sample and leave one blank. Add 200 µl of lysis reagent to each tube. Transfer 5 µl of enriched sample to the corresponding lysis tube. Add 5 µl uninoculated secondary enrichment broth as a DNA-free control.
5. Put tubes in a heating block at 55°C for 60 min.
6. Transfer the tubes to the second heating block at 95°C for 10 min.
7. Allow the lysate to cool 5 min in a cooling block or on ice before transferring to PCR tubes. Turn the thermocycler on at this time and set to 90°C hold.
8. Label the BAX tubes with a permanent marking pen. To apply DNA, transfer 50 µl of cooled lysed sample to a BAX tube. Make sure that the tubes are tightly covered with caps. Transfer the tubes to the thermocycler (prewarmed and held at 90°C). Start the BAX program in the thermal cycler. The machine will hold samples at 25°C when finished.

9. While samples are in the thermal cycler, prepare a 2% agarose gel (using nucleic acid grade agarose) in 0.5× Tris-borate-EDTA (TBE) buffer (44.5 mM Tris, 44.5 mM boric acid, 1 mM ethylene diamine tetra-acetic acid [EDTA] pH 8.3) with 0.22 μg/ml ethidium bromide. Alternatively, gels can be stained afterward in ethidium bromide or another fluorescent DNA stain.

10. Place the solidified gel into the apparatus and cover with 0.5× TBE.

11. Mix each BAX-amplified sample with 15 μl loading dye (supplied with kit) and load gel. Make sure that one lane contains the DNA ladder supplied with the kit. Run gels at 180V constant voltage for 25 min.

12. Visualize gels on a UV light transilluminator, and document gels with a digital or film camera.

    a. Use eye protection.

    b. Gloves should be worn whenever handling ethidium bromide.

13. If results are positive, use secondary enrichment to streak on Oxford agar, and confirm as described in Laboratory 11.

■ IV. RESULTS

Include a photograph or illustration of your agarose gel. Label size standards, *L. monocytogenes* bands, and control bands. Were any food samples positive for *L. monocytogenes*?

■ V. DISCUSSION QUESTIONS

1. Why is it important to perform culture biochemical confirmatory testing after having a positive result in a rapid test?

2. What is the purpose of the 200 bp internal positive control in the BAX system?

3. How could a PCR-based method be modified so that nucleic acid is only amplified from viable cells?

LABORATORY NOTES

◼◼ LABORATORY NOTES

■ LABORATORY NOTES

LABORATORY 13

ENUMERATION OF SPORES FROM PEPPER

■ I. OBJECTIVE

• Learn the basics of performing a general spore count.

■ II. BACKGROUND

Certain types of bacteria produce spores in response to environmental stresses. In general, bacterial spores are dormant forms of cells that are much more resistant to heat, dehydration, freezing, and irradiation than the vegetative forms of the cells. In food processing, these increased resistances are of concern for food spoilage and food safety reasons. Spores are associated in nature with soil, dust, and water. Some spore-forming organisms such as *Clostridium perfringens* are also associated with the intestinal tracts of humans and animals. Because spores have a high resistance to desiccation, it is not surprising that spores are detected in many foods containing dehydrated ingredients, such as starches, vegetables, grains, and spices.

Spore-forming bacteria are mainly from two genera of Gram-positive bacteria: *Clostridium* and *Bacillus*. Members of the *Clostridium* genus are strict (i.e., will only grow anaerobic) or facultative (prefers to grow in anaerobic) anaerobes. The members of the *Bacillus* genera tend to be aerobes or facultative anaerobes, and most produce the enzyme catalase. Spore-forming bacteria can also be divided by optimal growth temperature. There are both mesophilic and thermophilic spore-forming bacteria. The heat resistance of spores can vary. In any processed food, the intrinsic heat resistance of the spore, along with the pH, atmosphere, and processing and postprocessing holding temperatures, will determine which spores will germinate and grow.

Spices are produced all over the world and are notorious for having high populations of bacteria (both vegetative and spores), yeasts, and molds. The microbial populations of spices are probably mainly residents on the plant that survived drying and storage. In addition, most spices are collected in tropical areas by primitive methods and may be exposed to many contaminants before they are dry enough to halt microbial growth. They also may be stored in conditions that make them easy prey for insects, vermin, and dust. For example, black pepper is produced in India, Sri Lanka, Malaysia, Indonesia, and Brazil.

In the production of commercial products, it is important to have the lowest initial spore count possible before processing, in order to achieve the best microbial quality after processing. Spores can get into foods from a number of dry ingredients: starches, grains, and spices. In fact, the spore load of spices is so great that the majority of spices used in commercial food processing in the United States are irradiated to lower the spore and vegetative bacterial load. Table spices sold in supermarkets are usually not irradiated and will have high spore levels. Table spices generally are not considered a serious public health threat, even with the high microbial levels.

In this laboratory instruction, we will perform a spore count from ground pepper. Before plating, you will heat your sample to 80°C for 30 min.[1] This heating step will be performed for two reasons: (a) all vegetative cells will be destroyed to assure that we will only enumerate spores, and (b) "heat shock" enhances the outgrowth of most spores. The heating step is thought to activate an enzyme that starts the breakdown of the spore cortex or "outgrowth." You will incubate your plates at 35°C for mesophilic growth and under two atmospheres (anaerobic and aerobic). Basically, you will perform a mesophilic aerobic spore count and a mesophilic anaerobic spore count. If we were interested in a thermophilic spore count, we could incubate the plates at higher temperatures (50 to 55°C).

There are a number of methods used to grow anaerobic bacteria in the bacteriology laboratory. One method is to use an anaerobic chamber — an enclosed chamber that is flushed with nitrogen and other gases to create an anaerobic environment. There are also incubators that can be flushed with a variety of gases. We will use a GasPak™ system (Becton Dickinson and Company, Cockeysville, MD). This system consists of a GasPak jar, single-use gas generator envelopes, and catalyst pellets (palladium). To use this system, the petri plates are placed in the anaerobe jar. Depending upon the size of the jar, one to three generator envelopes may be needed to create an anaerobic environment. The generator envelopes should be cut open and placed in the jar in a vertical position. Add 10 ml of water to each, and then seal the jar. Once the jar is sealed, water in the envelope reacts with a sodium borohydride tablet to produce hydrogen that subsequently combines with the oxygen (in the presence of the palladium catalyst) to produce water (thus eliminating O_2 from the atmosphere). The generator envelope also consists of a sodium bicarbonate plus citric acid tablet, which in the presence of water produces CO_2. The presence of CO_2 stimulates the growth of some anaerobic bacteria. According to the manufacturer, after 2 h, the O_2 concentration will be reduced to less than 1%, and the CO_2 concentration will be between 4 to 10%. The entire jar system is placed into the incubator.

Bacteriological Media Used in this Lab

1. Tryptic soy agar with glucose (TSA + G): This is a nonselective and nondifferential media containing tryptone (digested casein) soytone (digested soybean meal) as a protein source, sodium chloride, and agar. Glucose (carbon source) is added to the level of 1% (10 g/liter).

■ III. METHODS

Class 1

Spore Count

Procedure

1. Weigh 1 g pepper in a plastic weigh bowl, and add it to 99 ml sterile peptone water. This will be an initial 1/100 dilution. Shake well.
2. Put the samples in an 80°C water bath for 30 min. This will heat shock spores and kill vegetative bacteria.

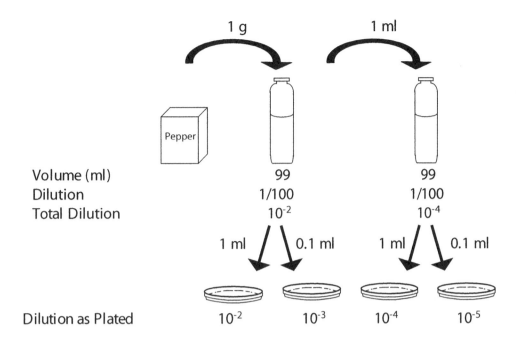

Figure 13.1 Dilution scheme for use in Laboratory 13.

3. Remove the bottle from the water bath. Mix well and wait for the pepper particulates to settle out of the solution.

4. Sample the liquid above the pepper particulates (because the particulates will clog your pipettes). Dilute and plate the heated pepper solution using a pour plate technique. Final plated dilutions should be 10^{-2} to 10^{-5} (Figure 13.1). We will be pour plating on TSA + G agar.

5. Plate each dilution in quadruplicate. Two of these plates will be incubated aerobically (A). The other two plates will be placed in an anaerobe jar and incubated anaerobically (AN). Make sure you label your group name and the atmosphere (A or AN).

Class 2

1. Count the CFU/plate, and record the results at each growth condition (35°C aerobic, 35°C anaerobic).

2. Calculate the CFU/g at each growth condition (35°C aerobic, 35°C anaerobic).

3. Record the colony morphologies of any surface growth.

■ IV. RESULTS

Mesophilic Aerobic Spore Count

Dilution as Plated	CFU/Plate	CFU/Plate	Average CFU/Plate
10^{-2}			
10^{-3}			
10^{-4}			
10^{-5}			

CFU/g =

Colony morphology observations:

Mesophilic Anaerobic Spore Count

Dilution as Plated	CFU/Plate	CFU/Plate	Average CFU/Plate
10^{-2}			
10^{-3}			
10^{-4}			
10^{-5}			

CFU/g =

Colony morphology observations:

▬ V. DISCUSSION QUESTIONS

1. If we had used spread plates, would you expect to see different colony morphologies and sizes with different growth conditions?

2. If you had a spore-forming bacterium that could grow under both aerobic and anaerobic conditions, how would you classify it, as a facultative or a strict anaerobe? Why?

3. Would you expect to have different results if we had not heated the pepper solution before plating?

4. If you were preparing a vacuum-packed processed meat product (e.g., bologna) that contained nitrites, would you be concerned about the addition of spices to the product? Why or why not?

■ LABORATORY NOTES

◼ LABORATORY NOTES

■ LABORATORY NOTES

LABORATORY 14

THERMAL DESTRUCTION OF MICROORGANISMS

■ I. OBJECTIVES

- Become familiar with heat process testing.

- Use group data to generate survival curves and calculate D values.

- Think about potential sources of laboratory error.

■ II. BACKGROUND

Heat is the most used method for inactivating microorganisms in food production. Microbial exposure to heat has two parameters: temperature and exposure time. In this laboratory, we will be exploring the heat destruction of nonpathogenic *Escherichia coli* K12. This organism is fairly heat sensitive, and destruction can be easily monitored in the laboratory.

To monitor cell destruction, high initial cell numbers (much higher than would be found in a food product) are often used to allow for detection of reduced numbers over a number of log reductions. For example, if you start with 5×10^4 CFU/ml (4 log), you can only detect a 2 to 3 log reduction (remember, 1×10^1 CFU/ml is below countable range), but if you start with 5×10^8 CFU/ml, you will have the opportunity to monitor up to a 6 to 7 log reduction.

In this lab, we will perform thermal destruction in standard test tubes in a water bath. This system is being used because it is standard equipment that can be found in most laboratories. However, it is not an ideal system to use to obtain accurate thermal destruction data. Ideally, the system should have a very rapid "come-up" time (the time to get to the internal processing temperature), and the entire processing vessel should be held at the same temperature so there is not a temperature gradient between the sample submerged in the water bath and the cap out of the water bath. For this reason, ideally, the bacterial suspension should be processed within sealed capillary tubes or vials that are completely submerged in a water or oil bath.

When you look up survival curves in your textbook, note that the survival curves are logarithmic in character, or there is a linear relationship between the logarithmic number of survivors and linear time. Although not usually discussed in textbooks, data generated during thermal destruction will often be sigmoidal in shape, meaning that there is an initial shoulder before the exponential drop in cell numbers or there is a tail at the end of a curve. The exact reasons behind shoulders and tails are not always obvious but are likely to be related to heterogeneity within the cultures. Recently, it was shown that thermal death curves can be related mathematically to the statistical properties of the underlying distribution of heat resistances within a culture.[25] In addition, the concept of "*D* values" has come under harsh criticism, because this concept assumes first-order destruction and ignores the shoulder and tailing values. However, D values and z values are still used by the U.S. Food and Drug Administration (FDA) to assure adequate thermal processing.

This lab will use the following to generate thermal destruction curves:

Organism: *Escherichia coli* K12 (nonpathogenic), stationary phase cells grown at 37°C with shaking (250 rpm)

Medium: Tryptic soy broth (TSB), pH 7.0

Processing temperature: 54, 58, or 62°C

Dilutions: 99 ml dilution blanks

Plating method: Pour plating in tryptic soy agar (TSA) with 1 and 0.1 ml plated volumes

Three different survival curves will be generated for a single organism, inoculum level, and growth condition.

■ III. METHODS

Class 1

Generate Data to Determine D Values
Procedure

1. Each group will be assigned one of the three temperatures (54, 58, or 62°C). Study dilutions given in Table 14.1 and the dilution scheme (Figure 14.1). At each time point, you will need to prepare only the dilutions given in Table 14.1. Organization is key for a timed experiment. Therefore, it is important to label all sterile dishes and dilution blanks needed for all time points before beginning any culture work. Make sure to include your name, processing temperature, time point, and dilutions as plated on the bottom of each plate.

2. Transfer 11 ml bacteria from overnight growth, to 99 ml test media (TSB pH 7), and mix well. [Note: Overnight growth has approximately 3×10^9 CFU/ml, so there will be an approximate initial bacterial level of 3×10^8 CFU/ml in the broth.]

3. Use a permanent marker (tape will fall off during heating) to label six sterile blank test tubes with your group name and time points (10, 20, 30, 40, 50, and 60 min).

4. Add 5 ml of the inoculated broth to each of the six tubes. These will be your 10^0 samples at each time point.

5. Add tubes labeled 10, 20, 30, 40, 50, and 60 min to your water bath. Start your timer counting up.

6. Plate your $T = 0$ sample from your original bottle of inoculated TSB. Plate final dilutions 10^{-5} to 10^{-7} in duplicate using pour plates.

7. At $T = 10$, remove your 10 min tube from the water bath, and place it in an ice or ice water bath to cool. Mix well and dilute and pour plate in duplicate according to Table 14.1.

8. At all subsequent time points, repeat step 7.

9. Incubate plates inverted at 37°C for 48 h.

Class 2

Results and Further Data Analysis

1. Record the raw data in the results section.

2. Plot the data using graph paper or a graphing program, such as Microsoft® Excel. Data can be plotted on a semi-log plot with surviving cell numbers (CFU/ml) on the logarithmic y-axis and

TABLE 14.1

Dilutions to Be Plated at Each Processing Temperature[a]

Sampling Time (min)	Final Plated Dilutions		
	54°C	58°C	62°C
0	$10^{-5}, 10^{-6}, 10^{-7}$	$10^{-5}, 10^{-6}, 10^{-7}$	$10^{-5}, 10^{-6}, 10^{-7}$
10	$10^{-5}, 10^{-6}, 10^{-7}$	$10^{-2}, 10^{-3}, 10^{-4}, 10^{-5}, 10^{-6}$	$10^{-1}, 10^{-2}, 10^{-3}, 10^{-4}, 10^{-5}$
20	$10^{-3}, 10^{-4}, 10^{-5}, 10^{-6}$	$10^{-1}, 10^{-2}, 10^{-3}, 10^{-4}, 10^{-5}$	$10^{-0}, 10^{-1}, 10^{-2}, 10^{-3}, 10^{-4}$
30	$10^{-3}, 10^{-4}, 10^{-5}, 10^{-6}$	$10^{-0}, 10^{-1}, 10^{-2}, 10^{-3}, 10^{-4}$	$10^{-0}, 10^{-1}, 10^{-2}, 10^{-3}$
40	$10^{-2}, 10^{-3}, 10^{-4}, 10^{-5}, 10^{-6}$	$10^{-0}, 10^{-1}, 10^{-2}, 10^{-3}$	$10^{-0}, 10^{-1}, 10^{-2}, 10^{-3}$
50	$10^{-2}, 10^{-3}, 10^{-4}, 10^{-5}$	$10^{-0}, 10^{-1}, 10^{-2}, 10^{-3}$	$10^{-0}, 10^{-1}, 10^{-2}$
60	$10^{-1}, 10^{-2}, 10^{-3}, 10^{-4}$	$10^{-0}, 10^{-1}, 10^{-2}, 10^{-3}$	$10^{-0}, 10^{-1}, 10^{-2}$

[a] These dilutions were optimized so that countable plates are obtained using *E. coli* K12. If another organism is used, dilutions may have to be broadened to assure countable plates.

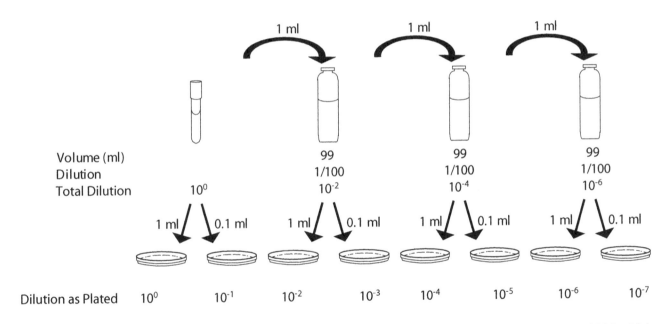

Figure 14.1 Dilution scheme for TDC of *E. coli* K12. Dilutions for each time point are given in Table 14.1.

time (in minutes) on the linear *x*-axis. This is known as the thermal destruction curve (TDC). Alternately, the cell levels (CFU/ml) can be converted to log cell numbers (log CFU/ml), and this can be plotted against time linearly.

3. Calculate the *D* value from the slope. The *D* value is defined as the amount of time required to reduce the bacterial population by 1 log (90% reduction) or the reciprocal of the thermal destruction curve slope. Calculation of *D* values assumes first-order destruction. Each group should record their *D* value for their temperature and provide a copy of their TDC slope to be distributed to the class. Use *D* values to calculate *z* values for each medium.

4. Plot *D* values vs. heating temperature on a second sheet of semi-log paper (provided). This is known as the thermal death time curve (TDT). The *z* value is defined as the change in temperature required for the thermal destruction curve to traverse one log cycle or the reciprocal of the slope of the TDC.

■ IV. RESULTS

Raw Data

Time (min)	Dilutions Plated and Counted	Calculated Surviving CFU/ml
0		
10		
20		
30		
40		
50		
60		

Calculated *D* Values and *z* Values

	Temperature		
	54°C	58°C	62°C
D value			
z Value			

■ V. DISCUSSION QUESTIONS

1. What types of experimental errors could be associated with this type of experiment? What could you do differently to eliminate these errors?

2. Why would a sealed tube that is completely submerged in the water bath yield more accurate results than using a test tube?

3. Were your thermal destruction curves straight lines? Assuming that the nonlinearity of your curves was not due to experimental error, what could have caused this variation?

4. If you wanted to repeat this experiment and look at the amount of injured cell levels, how could you perform that experiment? Would you need to adjust the dilutions that you plate? Why or why not?

5. Using the D values from your thermal destruction curve, how long would you have to heat your TSA to assure a $3D$ destruction? A $5D$ destruction? What if the pH of your medium was at pH 4? Would you expect to see differences? Why or why not?

6. Many people are criticizing the $12D$ concept, because it is pure extrapolation. Why would it be nearly impossible to experimentally confirm a $12D$ lethality?

LABORATORY NOTES

■ LABORATORY NOTES

■ LABORATORY NOTES

LABORATORY 15

CANNING AND SPOILAGE
OF LOW-ACID PRODUCTS

■ I. OBJECTIVE

- Learn about retort processing and spoilage of canned products.

■ II. BACKGROUND

Spore-forming bacteria of concern in foods are mainly from two genera of Gram-positive bacteria: *Clostridium* and *Bacillus*. Members of the *Clostridium* genus are strict (i.e., will only grow anaerobic) or facultative (prefer to grow in anaerobic) anaerobes. Members of the *Bacillus* genera are aerobic to facultatively anaerobic. A more detailed discussion is in the background information of Laboratory 14. Spore-forming bacteria can also be divided by optimal growth temperature. Within both genera, there are spore-forming organisms that grow as mesophiles or thermophiles. In addition, the heat resistance of spores can vary between species and will also be dependent upon food product composition.

There are a number of mesophilic anaerobic spore-forming bacteria that should be destroyed by retort processing and will not grow in high-acid foods (pH < 4.5). These organisms are members of the *Clostridium* genus. The presence of viable mesophilic anaerobic spores in a canned product indicates insufficient processing or the possibility of can leakage. In either case, there is the possibility that *Clostridium botulinum* may be present.

Mesophilic anaerobic spore-forming bacteria can be divided into two groups based upon their ability to digest complex proteins: (a) proteolytic (can digest protein) and (b) nonproteolytic. Another name for a proteolytic mesophilic anaerobic spore former is putrefactive anaerobe, because these organisms produce hydrogen sulfite along with gases such as ammonia (from the breakdown of proteins), carbon dioxide, and hydrogen. Other mesophilic *Clostridium* spp. are nonproteolytic and can grow anaerobically without producing gas (no bulging of can). *C. botulinum* falls within this mesophilic anaerobic spore-forming bacteria group: some strains are proteolytic (can produce gas), while other strains are nonproteolytic (no gas is produced).

The low-acid canning process is designed to eliminate *C. botulinum*. The U.S. government classifies *C. botulinum* as a biological hazard level 3. This means that research with this organism is required to be performed in a strict containment environment. Therefore, spores from *C. botulinum* are rarely used for testing thermal or nonthermal processes. Instead, spores from a nonpathogenic *Clostridium* spp., such as *Clostridium sporogenes* PA3679 (ATCC 7955), are used. The growth characteristics of *C. sporogenes* and *C. botulinum* are similar in that both are mesophilic anaerobes; however, spores of

C. sporogenes are slightly more heat resistant ($D_{121}°C$ = 0.5 to 1.5 min) than those of *C. botulinum* ($D_{121}°C$ = 0.1 to 0.2 min).[14]

Even when cans are processed to a theoretical "12*D* kill" of *C. botulinum*, there are a number of spore types that are still viable. This is why canned foods are called "commercially sterile." These highly heat-resistant spore-forming organisms are considered to be nonpathogenic spoilage organisms and will only grow if the can is stored at high temperatures. If the canning process temperature and time were increased to kill these organisms, both the textural and nutritional qualities of the product would greatly decrease. Two things can be done to prevent spoilage of commercially canned products by retort-resistant spores: have the lowest number of spores as you can in the product before canning, and store the canned product below 40°C because most retort-resistant spores are thermophilic. Examples of thermophilic spore-forming spoilage bacteria are *Bacillus stearothermophilus* (a facultative anaerobe, a "flat-sour" spoilage organism), *Clostridium thermosaccharolyticum* (a facultative anaerobe that produces abundant gas from sugars), and *Desulfotomaculum nigrificans* (a facultative anaerobe that produces a strong odor of hydrogen sulfite without gas).

In this lab, you will inoculate commercially canned corn with spores, prior to heat processing, for various amounts of time. We will use a commercially canned product to assure that *C. botulinum* spores were destroyed. Two spore-forming organisms will be used:

C. sporogenes ATCC 7955 (PA3679): The "PA" in PA3679 stands for putrefactive anaerobe. These spores have a *D* value just higher than *C. botulinum* ($D_{121}°C$ = 0.1 to 1.5).[2] If viable spores are present in a low-acid (pH > 4.6) canned food, this organism is one of several *Clostridium* spp. (including proteolytic *C. botulinum*) that can cause putrefactive anaerobic spoilage, which is indicated by the swelling and sometimes bursting of cans and a characteristic putrid stench. Spores of this strain are often used in commercial thermal challenge studies in place of *C. botulinum*.[18]

B. stearothermophilus: This is a thermophilic facultative anaerobic spore former that is associated with "flat-sour spoilage" of both low-acid and acidic foods. Spores of this organism have a $D_{121}°C$ of 1 to 5.8 min, depending upon the suspending medium.[2] Spoilage by this organism does not cause cans to bulge (characterized as "flat") but often causes the pH of the product to drop (characterized as "sour").[26] This organism mainly grows at elevated temperatures, with an optimal growth temperature of 60°C. Therefore, most bacteriological analyses for thermophilic spore formers are performed at 55°C.

■ III. METHODS

Class 1

Inoculation of Corn and Retort Processing
Procedure

1. Aseptically open the can of a commercially canned corn. The purpose behind this is to assure that *C. botulinum* spores were destroyed:

   a. Sanitize with freshly prepared 100 ppm bleach solution, and allow it to sit on a surface for 10 min.

   b. Pour off excess sanitizing agent, and heat the lid with a burner flame until the moisture evaporates.

   c. Cut a hole in the lid with a sanitized Bacti® Disk Cutter (if available) or an autoclaved metal can opener.

2. Pour the contents of the can into a sterile beaker:

   a. Use a sterile utensil and take 25 g sample of corn and brine for pH measurement.

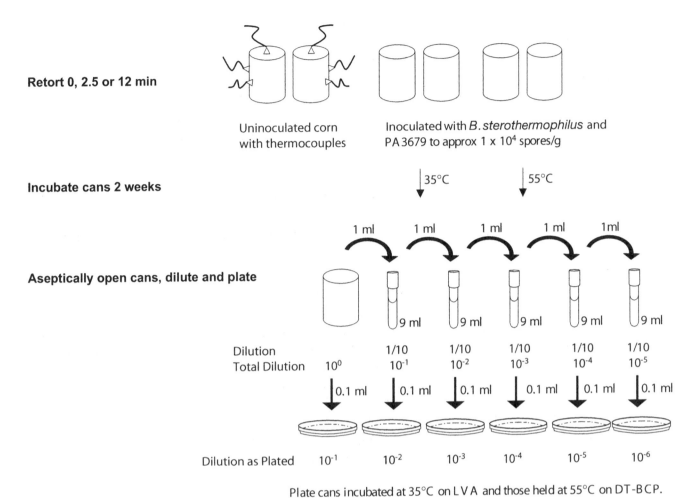

Retort 0, 2.5 or 12 min

Uninoculated corn
with thermocouples

Inoculated with *B. sterothermophilus* and
PA 3679 to approx 1 x 10⁴ spores/g

Incubate cans 2 weeks

35°C   55°C

Aseptically open cans, dilute and plate

1 ml   1 ml   1 ml   1 ml   1ml

9 ml   9 ml   9 ml   9 ml   9 ml

| Dilution | | 1/10 | 1/10 | 1/10 | 1/10 | 1/10 |
| Total Dilution | 10^0 | 10^{-1} | 10^{-2} | 10^{-3} | 10^{-4} | 10^{-5} |

0.1 ml   0.1 ml   0.1 ml   0.1 ml   0.1 ml   0.1 ml

| Dilution as Plated | 10^{-1} | 10^{-2} | 10^{-3} | 10^{-4} | 10^{-5} | 10^{-6} |

Plate cans incubated at 35°C on LVA and those held at 55°C on DT-BCP.

Figure 15.1 Flowchart of procedures and suggested dilution scheme.

b. Pack four control cans outfitted with thermal couples with uninoculated corn. These will be the two control cans used to monitor the processing temperature profile for each processing time.

c. Use a sterile utensil and take 25 g sample of noninoculated corn to use for the microbiological analysis of the control. Dilute to 10^{-2} (Figure 15.1):

i. To determine background levels of mesophilic anaerobes, plate to final dilution of 10^{-1} to 10^{-3} on veal–liver agar (VLA), and incubate anaerobically at 35°C for 48 h.

ii. To determine the background levels of flat-sour spore-forming organisms, plate to final dilutions of 10^{-1} to 10^{-3} on glucose tryptone bromcresol purple agar (GT-BCP) at 55°C aerobically for 48 h. Flat-sour spore formers will form yellow colonies on GT-BCP.

d. Inoculate with spore stocks[27,28]:

i. Weigh the remaining corn and brine. Use this weight to calculate the amount of spore stock to add. The amount added will depend on the weight of your sample and the number of viable spores present in your spore stocks.

ii. Inoculate the remaining corn and brine with *C. sporogenes* ATCC 7955 (PA3679) and *B. stearothermophilus* to the final level of approximately 1×10^4 CFU/g of each organism.

iii. Mix thoroughly with a sterile spoon.

e. To determine the actual level of each spore type in the sample, remove a 25 ml sample of brine and place it in a sterile screw cap test tube. Place at 75°C for 20 min to heat shock and dilute and plate and to determine the actual level of each spore type:

   i. To determine background levels of PA3679 spores, dilute and plate 10^{-1} to 10^{-4} (Figure 15.1) on VLA, and incubate anaerobically at 35°C for 48 h.

   ii. To determine background levels of *B. stearothermophilus* spores, dilute and plate 10^{-1} to 10^{-4} (Figure 15.1) on GT-BCP at 55°C aerobically for 48 h. Colonies will appear yellow on GT-BCP.

3. Canning:

   a. Use a 211 × 300 (2 11/16 in diameter × 3 in height) size can or other appropriate small can. A 211 × 300 can holds approximately 7 oz.

   b. Fill eight cans to the top with inoculated corn and brine. Clean any spills on the rim with sterile cloth.

   c. Seal using a can sealer.

4. Do not retort four cans. These will be the zero-processing time controls.

5. Retort at 121°C for 2.5 and 12 min. This will need to be performed in batches consisting of four inoculated cans and two uninoculated cans with thermal couples. Timing should not start until the coldest point of the control cans reached the target temperature. Because this is a batch process, plan for at least an additional 40 min per run for the "come-up" and "cool-down" times. (If a retort is not available, an autoclave or pressure cooker can be used. However, the internal temperature of the cans cannot be monitored.) As the cans are removed from retort, be sure to label them with the appropriate retort times.

6. After processing, allow the cans to cool. Wrap the cans in plastic bags (in case they burst). Incubate two cans from each processing time (0, 2.5, and 12 min) at 35°C and 55°C.

2-Week Incubation

Check the cans daily and record the results. Terms to describe cans include the following:

Flat: A can with standard flat ends.[29]

Flipper: A can that appears flat, but when it is "brought down sharply on its end on a flat surface, one end flips out. When pressure is applied to this end, it flips in again and appears flat."[29]

Springer: A can with one end permanently bulged. If enough pressure is applied, the end will push in and the opposite end will pop out.[29]

Soft swell: A can with bulges at both ends. When thumb pressure is applied, the ends can be pushed in.[29]

Hard swell: A can with bulges at both ends. When thumb pressure is applied, the ends cannot be pushed in.[29] Buckling of the can is usually observed before bursting. When the can bursts, it will likely burst where the double seam meets the side seam.

Class 2 (2 Weeks Later)

1. After 2 weeks, pull the cans from the incubator and open the cans aseptically (as described above):

   a. If the cans are bulging, chill the cans in the refrigerator before opening and do not flame to remove sanitizer. Cover with a sterile towel while opening, in case a portion of the contents spray out.

2. Remove a portion of the brine to check the pH.

3. Dilute and plate the brine according to Figure 15.1. This time, plate the entire dilution scheme, because there may be few organisms or large numbers:

    a. Dilute and plate cans held at 35°C on VLA. Incubate VLA at 37°C anaerobically for 48 h to determine the number of viable *C. sporogenes*.

    b. Dilute and plate cans held at 55°C onto dextrose tryptone bromcresol purple agar (DT-BCP), and incubate aerobically at 55°C for 48 h to detect viable *B. stearothermophilus*.

■ IV. RESULTS

Initial pH of corn brine =

Background Levels in Corn
Mesophilic Anaerobic Spore Formers

Dilution as Plates	CFU/Plate	CFU/Plate	Average CFU/Plate
10^{-1}			
10^{-2}			
10^{-3}			
10^{-4}			
10^{-5}			

Background level CFU/ml =

Flat-Sour Spore Formers

Dilution as Plates	CFU/Plate	CFU/Plate	Average CFU/Plate
10^{-1}			
10^{-2}			
10^{-3}			
10^{-4}			
10^{-5}			

Background level CFU/ml =

Initial Spore Levels in Inoculated Samples

C. sporogenes PA3679

Dilution as Plates	CFU/Plate	CFU/Plate	Average CFU/Plate
10^{-1}			
10^{-2}			
10^{-3}			
10^{-4}			
10^{-5}			

CFU/ml =

B. stearothermophilus

Dilution as Plates	CFU/Plate	CFU/Plate	Average CFU/Plate
10^{-1}			
10^{-2}			
10^{-3}			
10^{-4}			
10^{-5}			

CFU/ml =

Describe cans incubated at 35°C and 55°C (include time and description: flat, bloated etc.).

Final Levels after Incubation for 2 Weeks
Mesophilic Anaerobic Spore Formers (35°C)

Dilution as Plates	CFU/Plate	CFU/Plate	Average CFU/Plate
10^{-1}			
10^{-2}			
10^{-3}			
10^{-4}			
10^{-5}			
10^{-6}			

CFU/ml =

pH of brine =

Flat-Sour Spore Formers (55°C)

Dilution as Plates	CFU/Plate	CFU/Plate	Average CFU/Plate
10^{-1}			
10^{-2}			
10^{-3}			
10^{-4}			
10^{-5}			
10^{-6}			

CFU/ml =

pH of brine =

◼ V. DISCUSSION QUESTIONS

1. Why are low-acid canned foods called "commercially sterile?"

2. Did you expect to find any mesophilic anaerobic spore formers or flat-sour spore formers in the commercially canned corn? Why or why not?

3. Why did you heat the inoculated brine sample at 75°C for 20 min prior to plating? Would your results have been different if you had not heated it?

4. Is the title "putrefactive anaerobe" an appropriate term with which to describe *C. sporogenes* PA3679?

5. How would changing the food item possibly affect the results? Discuss differences in solids contents of various types of low-acid canned foods.

LABORATORY NOTES

■ LABORATORY NOTES

■ LABORATORY NOTES

LABORATORY 16

COMBINED EFFECTS OF INTRINSIC FORMULATION AND EXTRINSIC FACTORS USING GRADIENT PLATES

I. OBJECTIVE

- Evaluate the combined effects of intrinsic media formulation and extrinsic factors on microbial growth.

II. BACKGROUND

There are intrinsic factors (pH, a_w, antimicrobials) and extrinsic factors (temperature, atmosphere) of foods that affect the growth of microorganisms. In the laboratory, traditionally, these factors were studied individually. The current trend in food preservation is "hurdle technology." In hurdle technology, multiple intrinsic and extrinsic factors are used together to create a synergistic barrier to prevent microbial growth. One factor that makes evaluating hurdle technology difficult is that the results of combinations of factors are unpredictable. In addition, the testing of multiple factors often involves complex preparation of many different media formulations.

Thomas and Winpenny published a paper in which they used a two-dimensional gradient plating scheme (similar to those commonly used in environmental microbiology) to monitor the effect of combined factors upon microbial growth.[30] In this lab, a simplified version of their methodology will be used to evaluate the intrinsic parameters of pH and NaCl concentration (a_w) and the extrinsic factors of growth temperature and atmosphere. It is important to note that this is a dynamic system, and these gradients are constantly changing. Therefore, gradient plates should only be used for initial screening, and subsequent experiments must be performed to evaluate growth within specific formulations of bacteriological media and food systems.

III. METHODS

Plate Preparation

Gradient plates should be prepared the day before by the instructor or a selected group of students. The gradient plates consist of four 15 ml layers. All layers will consist of brain–heart infusion broth enriched with yeast extract (0.3%), glucose (0.3%) (BHIYEG), and agar (1.5%).

The gradient plates were prepared as follows:

Acid layer (pH 3.5): This is the bottom layer that was poured with one end of the plate elevated to form an agar wedge. The acidified broth (2× BHIYEG pH 3.5) and the agar are prepared separately, because the combination of high heat in the autoclave and low pH accelerate the hydrolysis of agar, and it would no longer solidify. Lactic acid was used to acidify the media.

Alkali layer (pH 10): This layer is poured on top of the acid layer wedge with the plate sitting level on the lab bench. Thus, the alkali layer fills in a complementary agar wedge to the acid agar wedge. This layer is prepared as 1× BHIYEG, and the pH is adjusted to 10 with the addition of NaOH. Agar (1.5%) is added, and the media is autoclaved, tempered, and added.

NaCl layer (15% w/v): This is the third layer. The plate is turned 180° and elevated to form an agar wedge. The salt gradient runs perpendicular to the acid gradient. Sodium chloride (15% w/v) is added to BHIYEG with agar (1.5%) prior to autoclaving. Media should be tempered prior to pouring plates.

BHIYEG: The final layer is comprised of BHIYEG with 1.5% agar. It is poured with the plate sitting flat on the lab bench. This layer forms the complementary wedge to the NaCl layer, thus finishing the salt gradient.

The plates should sit overnight at room temperature before starting the measurement and inoculation portion of the experiment. These gradients are dynamic (constantly changing). It is important not to let the plates sit too long, because the pH and the salt gradient could eventually reach equilibrium.

Class 1

Measurement of Gradients
Each group should measure both the pH and NaCl gradient from one of their plates. To do this, the pH should be measured first and the NaCl gradient second.

Measurement of Acidity Gradient
Ideally, a flat-bottom electrode should be used to measure this gradient:

1. Calibrate the pH electrode using standard buffers at pH 7.0 and 4.0.
2. At the top center of the plate (such as square A4), place the electrode on the surface and wait for a steady reading. Record the value.
3. Move the electrode to the next 12 mm square across the pH gradient (such as square B4) and record the pH. You should obtain six readings. You can record your readings on Figure 16.1A in the results section.

Measurement of Salt Gradient
To measure the amount of salt in the gradient, we will be using a conductivity meter:

1. Remove the agar from all four layers using a cork bore. The sampling is one of the trickiest parts of this lab, but (thankfully) this analysis does not need to be performed under aseptic conditions.
2. Start at one end on the salt gradient near the middle of the plate (such as square 1D). Take the cork bore (or the large end of a Pasteur pipette) and press into the center of the square until it hits the bottom of the plate.
3. While turning it, bring up the cork bore out of the agar. All four layers should be within the cork bore. Use a cool inoculating needle, and release the sample into a test tube containing 10 ml of distilled water. If all or some layers remain in the plate, an inoculating needle or flat-tipped forceps can be used to carefully dig it out. Be careful not to crush the sample. For accuracy, it is important that all samples be the same size.

4. Continue sampling across the salt gradient taking one core sample from each row (such as 1D, 2D, 3D, etc.). Each group should have six samples.

5. Melt the agar by placing the tubes in a boiling water bath over the flame on your bench. Allow samples to cool.

6. Use a conductivity meter to measure the conductivity of the samples.

7. Use the standard curve to correlate your conductivity readings to the concentration gradient supplied by your instructor. Record your results in Figure 16.1B.

Plate Inoculation

A variety of organisms can be used for this experiment and will be supplied by your instructor. The purpose of this inoculation is to use enough organisms so that under permissive conditions they grow into a confluent lawn. For this reason, overnight broth cultures will be used without dilutions:

1. Each group should inoculate the remaining four plates with the same organism.

2. Inoculate each plate with 0.25 ml of overnight growth of your target organism, and spread with a flamed "hockey stick."

3. Incubate one plate at 30°C, one plate at 30°C in an anaerobic chamber, one plate at 37°C, and one plate at 37°C in an anaerobic chamber.

4. Incubate for 48 h.

Class 2

Results of Lawn Growth on Gradient Plates

Study each plate. Visible growth indicates regions where the organisms have the ability to grow. Regions that do not have growth may indicate a "bacteriostatic" region (not growing, but not dead) or a bacteriocidal region (bacterial killing) on the plate. Record the results by drawing (in Figure 16.1C through Figure 16.1F) or by using a digital camera. Compare these results to the results obtained during measurement of the pH and NaCl levels. If we had time, we could evaluate bacteriocidal vs. bacteriostatic regions by duplicate plating. To do this, we would take sterile velvet and press it to our gradient plate. This could then be transferred to a plate without any inhibitory reagents, and we could watch for growth.

■■■ IV. RESULTS

A. Measured pH gradient

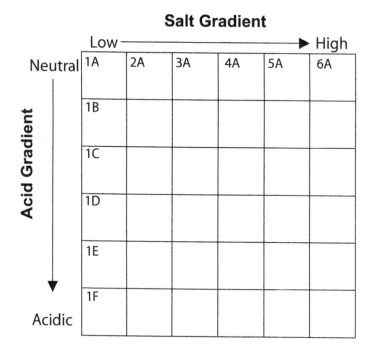

B. Measured Salt Gradient

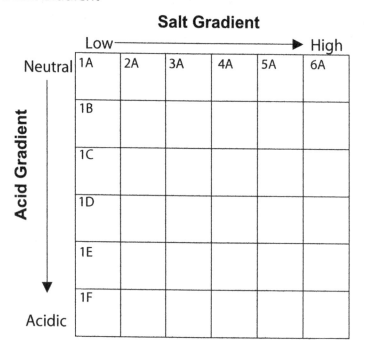

Figure 16.1 Measurement of pH and salt gradients (A, B). Drawings of petri plate growth (C through F).

C. Incubation Temp_____ Atmosphere_____

D. Incubation Temp_____ Atmosphere_____

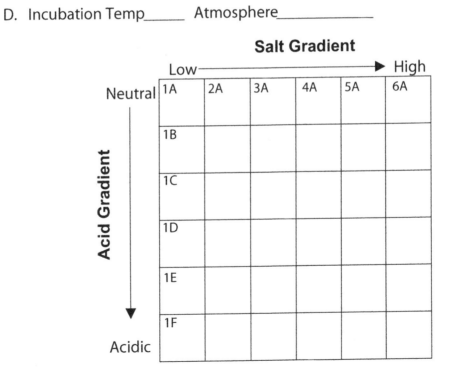

Figure 16.1 (continued)

E. Incubation Temp_____ Atmosphere_____

Salt Gradient

Salt Gradient

F. Incubation Temp_____ Atmosphere_____

Salt Gradient

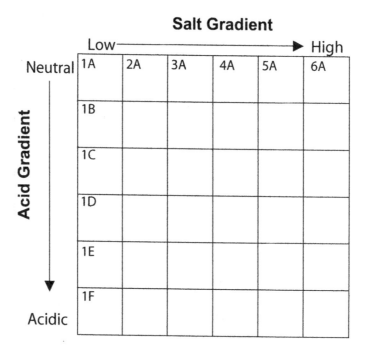

Figure 16.1 (continued)

◼ V. DISCUSSION QUESTIONS

1. What errors are associated with the gradient plate technique? If you only wanted to study the growth of an organism in a narrow range of pH and a_w, how would you do it?

2. Was the growth pattern you observed the same at both temperatures? Was the growth pattern the same under aerobic and anaerobic incubation?

3. If you performed the same experiment using glycerol to lower the a_w, would you expect the results to be different? What about if you used an inorganic acid to lower the pH?

4. Look at the pH and salt content of the plates that had no growth under any of the extrinsic conditions (temperature and atmosphere). Are there any foods that might have similar pH and salt contents? Are these levels realistic for inhibiting bacterial growth in a good-tasting food product?

■ LABORATORY NOTES

■■■ LABORATORY NOTES

■ LABORATORY NOTES

LABORATORY 17

CLEANING AND SANITATION

■ I. OBJECTIVES

- Understand the basics of cleaning and sanitation.

- Learn methods of surface sampling.

- Become familiar with the adenosine triphosphate (ATP) luciferase method.

■ II. BACKGROUND

Cleaning and sanitation are critical for safe food handling in all aspects. Whether it occurs in the food manufacturing plant, commercial kitchen, or in the home, the ability to eliminate bacteria from food contact surfaces is important to prevent contamination of food products.

There are a number of ways in which sanitation should be validated. First, the level of sanitizing agent should be confirmed using testing kits. This is important because many sanitizing agents are actually less effective at high concentrations. Second, the appropriate types of cleaning agent and sanitizer should be selected for each part of the processing plant, because some cleaning and sanitizing agents are only active in a certain pH range or can be neutralized by food components (i.e., protein or fats). Third, effectiveness in removing and destroying bacteria from the cleaned and sanitized surfaces should be tested. Confirmation of sanitation effectiveness can be performed by direct contact with agar surfaces or by using wet swabs or sponges to sample surface bacteria, after which the bacteria are physically removed and detected by plating. A third method measures the amount of ATP present after cleaning and sanitizing, which is an indication of remaining viable bacteria or food residues.

Food sanitation involves both cleaning and sanitation.[31] Cleaning is performed to remove soil (dirt, dust, and organic matter). Sanitation is a treatment designed to eliminate bacterial, mold, yeast, and (in some cases) viral contamination. Cleaning and sanitizing go hand in hand, because soils can inactivate sanitizing agents.

The basic steps for cleaning and sanitizing are as follows:

1. Prerinse to remove particulates
2. Clean to remove soil
3. Rinse
4. Sanitize
5. (Rinse, if required)

In this lab, you will evaluate some cleaning and sanitizing agents used in the home. Our testing today will be basic so as to detect reductions in food or environmental contamination. The Association of Analytical Chemists (AOAC) has official protocols for testing these chemicals using large numbers of known pathogens. Our testing will be less stringent but should give us some idea about the usefulness of sanitizing agents used in the home.

Sanitizing agents can have two different types of an effect on microorganisms: they can be bacteriocidal or bacteriostatic. Bacteriocidal means that the chemical kills the organism, while bacteriostatic compounds will just prevent bacterial growth. Therefore, sampling methods that have plating steps utilize neutralization media designed to neutralize sanitizing agents (such as QUATs, phenol, iodine, chlorine, formaldehyde, and gluteraldehyde). If bacteria survive a sanitizing treatment, the buffer or agar will neutralize any sanitizing agent residuals so that the microorganisms can grow.

For testing methods, we will be using three methods:

- Agar contact: For this portion a Hycheck™ contact slide will be used to evaluate the microbiological contamination of surfaces. The yellow side contains plate count agar (PCA), while the purple side contains neutralization agar. We are going to use the PCA side to evaluate the microbiological load before cleaning and sanitation and the neutralization side to evaluate the same area after sanitation.

- Swabbing and plating: This is the most common type of surface evaluation. It is used to evaluate both processing surfaces and surfaces of foods (such as beef carcasses in a slaughtering plant). Sampling of bacteria involves rubbing a known surface area (sterile templates can be used) with a sterile cotton swab or sponge wetted with a sampling buffer. If the surface has residues of cleaning and sanitizing agents, a neutralizing buffer is used to eliminate possible bacteriostatic effects. After sampling, the swab or sponge is placed in a known volume of buffer, and the bacteria are removed from the swab and sponge by vortexing or Stomacher homogenizer, respectively. The buffer is then diluted and plated to assess the level of CFU/cm^2. If you need to determine the presence or absence of a pathogen, the swab or sponge can be added to a selective enrichment broth, and isolation can proceed.

- Adenosine triphosphate (ATP) luciferase: ATP is found in all living cells. It is the universal agent used for transfer of energy. ATP is found in live cells of plant and animal origins as well as in live bacterial cells. ATP is present at a concentration of 10^{-18} to 10^{-17} mole/bacterial cells and disappears within 2 h after bacterial death.[18] ATP tests can be used to rapidly evaluate the effectiveness of cleaning and sanitation in a food-processing plant. The ATP present may be due to food residues or to the presence of viable bacteria, but in either case, the presence of ATP on food-processing surfaces after cleaning and sanitation is an indication of poor hygiene:

  - The most common ATP test uses an enzyme called luciferase and a substrate called luciferin. This is the same system used by fireflies to generate their flashes of light. This test is fast (takes just a few seconds) and sensitive (can detect as little as 10^1 to 10^3 fg [or 10^{-14} to 10^{-12} g] of ATP.[14,18] The test works like this:

    - Swab a surface to pick up bacteria or other ATP sources.
    - Add the bacteria to the reaction mixture in the presence of oxygen (O_2). If ATP is present, the following reaction outlined in Figure 17.1 will occur.
    - Measure the amount of light. The amount of light present is proportional to the amount of ATP present on your food-processing surface.

  - The light is measured in an instrument called a luminometer that measures light and reports the value in relative light units (RLUs). In addition, many instruments for use with commercial ATP swabs will report a value of pass or fail. The instrument is programmed to subtract a

$$LUCIFERIN + LUCIFERASE + ATP \longrightarrow$$

$$LUCIFERIN \cdot LUCIFERASE \cdot AMP \, (complex) + PP$$

$$LUCIFERIN \cdot LUCIFERASE \cdot AMP \, (complex) + O_2 \longrightarrow$$

$$OXYLUCIFERIN + LUCIFERASE + AMP + CO_2$$
$$+LIGHT$$

Figure 17.1 Detection of ATP using the luciferase–luciferin reaction. (Adapted from Adams, M.R. and Moss, M.O., *Food Microbiology*, 2nd ed., The Royal Society of Chemistry, Cambridge, 2000; Banwart, G.J., *Basic Food Microbiology*, 2nd ed., Chapman and Hall, New York, 1989.)

predetermined background level plus three times the standard deviation sample from each sample. Any value under the background with three standard deviations passes, while swabs with values three standard deviations above the background level fail.

■ III. METHODS

Agar Contact

Procedure

1. Each group will get one Hycheck contact slide.
2. Find an area around the building, such as a water fountain, bathroom, or doorknob. Note the location in your notebook.
3. Unscrew the top and remove the contact slide by holding the cap. Make sure you do not touch the agar surface.
4. Hold the spike on the bottom of the panel against the surface to be tested. Press down to bend the paddle around the hinge line. Gently lower the PCA (yellow) portion of the slide, and press the agar into contact with the test surface, still holding the slide by the cap.
5. Apply firm and even pressure on the test surface for a few seconds. Take care not to move the slide at this time, and then gently move the slide off the surface.
6. Place the slide back into the tube.
7. Clean and sanitize the area with a home cleaning agent according to the directions. If you were not lucky enough to obtain a commercial home sanitizing agent, use 500 ppm bleach solution from the lab and wipe with a paper towel.
8. Test the microbial load after cleaning with the neutralizing agar (purple) side of the same slide.
9. Place the slide back into the tube and incubate at 35°C for 48 h.
10. Count the total numbers of CFU on each side of the Hycheck contact slide. Divide this number by the slide area (5 cm²). Report the results as CFU/cm² pre- and postsanitizing.

Evaluation of Cleaning/Sanitizing Using Swab and Plate and ATP Detection

For this portion of the laboratory, we will sample a meat contact surface before and after washing and sanitizing.

Procedure

1. With a sterile tongue depressor, take some ground beef and spread it around in a 12 cm² square petri dish. This will be our simulated food preparation surface.

2. Remove most of the visible meat from the tray using the same tongue depressor.

3. Take a sterile swab. Moisten it in 9 ml neutralization broth, and press the swab on the inside of the test tube to wring out excess broth. The swab should be moist but not dripping wet.

4. The square petri dish is divided into 36, 1 cm² sections. Rub the swab over 9 cm² (¼ of plate area) within the petri dish.

5. Place swab back in 9 ml buffer. Label the tube as a reminder that this is precleaning and sanitation.

6. Swab the remaining squares with the Charm PocketSwab™. Pull the swab out of the top chamber of the PocketSwab. Be careful not to touch anything with this swab. Swab a different 9 cm² of the plate. Place the swab back into the PocketSwab, but do not twist it down. Label as precleaning and sanitation.

7. Clean and sanitize the tray at the sink. Choose one of the following methods and be sure to write it in your notebook:

   a. Select any combination of cleaning and sanitation agents, and use the five steps of cleaning and sanitizing (rinse, clean, rinse, sanitize, rinse).

   b. Use a standard dishwashing liquid, and treat the surface as you would in your own kitchen.

8. Repeat steps 3 through 6 with a second sterile swab moistened from a second tube of neutralization broth. Make sure you swab the same nine squares of the plate as you did in step 4. In addition, use a PocketSwab to sample from the remaining squares. Make sure that the second tube of neutralization broth and PocketSwab are labeled "postcleaning/sanitation."

9. The luciferase reaction occurs quickly. Therefore, these steps should be done in front of the luminometer, and each swab should be added to the luciferase mixture just before reading:

   • Screw the swab down all the way through the second membrane into the reaction mixture.

   • Tap the tube one to two times to mix.

   • Unscrew to remove the swab from the clear tube.

   • Tap once more.

   • Place the reaction in the luminometer.

   • Record RLU and pass or fail for your pre- and postcleaning and sanitation samples.

10. To isolate cells from the swab in the neutralization broth (pre- and postcleaning), vortex each tube at the highest speed for 2 min to release microorganisms from the swab before diluting. Dilute in saline and pour plate swab samples from before and after sanitizing, according to the dilution scheme (Figure 17.2). Each dilution should be plated in duplicate. Make sure you label your plates, pre- and postcleaning/sanitation. Each group should have 12 plates in the end: six from precleaning and six from postcleaning/sanitation. Incubate at 35°C for 48 h. Convert the results from CFU/ml to CFU/cm².

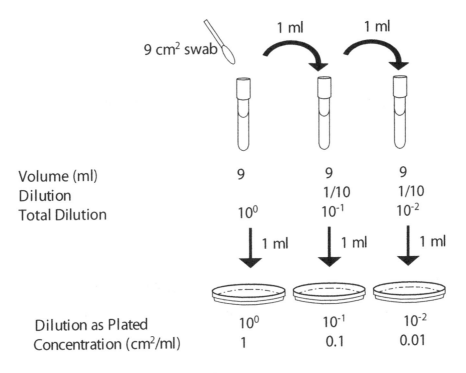

Figure 17.2 Dilution scheme for plating from swab.

■ IV. RESULTS

Hycheck Contact Slide Results

Before Cleaning (CFU/cm²)	After Cleaning and Sanitizing (CFU/cm²)

Cell Numbers Before and After Cleaning and Sanitizing Beef Contact Plate

Dilution as Plated	CFU/Plate	CFU/Plate	Average CFU/Plate	CFU/cm²
Before Cleaning				
10^0				
10^{-1}				
10^{-2}				
After Cleaning				
10^0				
10^{-1}				
10^{-2}				

ATP Levels from Ground Beef Contact Plates Before and After Cleaning and Sanitizing

Before Cleaning (RLU/9 cm²)	After Cleaning and Sanitizing (RLU/9 cm²)

■ V. DISCUSSION QUESTIONS

1. Why is cleaning important before sanitizing?

2. What are some advantages and disadvantages of ATP luciferase measurement for confirmation of sanitation?

3. Did your results surprise you? How did it come out differently than you expected?

■ LABORATORY NOTES

▰ LABORATORY NOTES

LABORATORY NOTES

LABORATORY 18

LUCIFERIN/LUCIFERASE DETECTION OF ATP ASSOCIATED WITH BACTERIA AND FOOD RESIDUES

I. OBJECTIVES

- Become familiar with rapid ATP luciferase detection methods.

- Evaluate the sensitivity of the ATP swabs using purified ATP.

- Compare standard ATP curve to levels of *Escherichia coli* K12 and food residues.

II. BACKGROUND

Adenosine triphosphate (ATP) is found in all living cells. It is the universal agent used for the transfer of energy. ATP is found in cells of plant and animal origin as well as in live bacterial cells. ATP is present at a concentration of 10^{-18} to 10^{-17} mole/bacterial cells and disappears within 2 h after bacterial death.[18]

ATP tests can be used to rapidly evaluate the effectiveness of cleaning and sanitation in a food-processing plant. The ATP present may be due to food residues or to the presence of viable bacteria, but in either case, the presence of ATP on food-processing surfaces after cleaning and sanitizing is an indication of poor hygiene.

The most common ATP test uses an enzyme called luciferase and a substrate called luciferin. In the presence of ATP and oxygen, this enzyme and substrate combination produces light and is the same system used by fireflies. This test is fast (takes just a few seconds) as well as sensitive (can detect as little as 10^1 to 10^3 femtomole [fmol] or 10^{-14} to 10^{-12} g of ATP under optimal conditions).

The test works like this:

- Swab a surface to pick up bacteria or other ATP sources.

- Add the sample to the luciferase–luciferin mixture.

- If ATP is present, in the presence of oxygen (O_2) the reaction outlined in Figure 18.1 will occur. The amount of light is measured and is proportional to the amount of ATP present on the swabbed sample.

LUCIFERIN + LUCIFERASE + ATP $\longrightarrow$

LUCIFERIN·LUCIFERASE·AMP (complex) + PP

LUCIFERIN·LUCIFERASE·AMP (complex) + O_2 $\longrightarrow$

OXYLUCIFERIN + LUCIFERASE + AMP + CO_2
+ LIGHT

Figure 18.1 [32] Luciferase/luciferin reaction in the presence of ATP (Adapted from Adams, M.R. and Moss, M.O., *Food Microbiology*, 2nd ed., The Royal Society of Chemistry, Cambridge, 2000; Banwart, G.J., *Basic Food Microbiology*, 2nd ed., Chapman and Hall, New York, 1989.)

All rapid ATP luminescence systems use a luminometer to measure light emitted. The values are reported as relative light units (RLU). In addition, some machines are programmed to give a report of pass or fail. For example, luminometers manufactured by Charm Sciences Inc. (Lawrence, MA) are programmed to subtract the swab background level plus three times the standard deviation from the displayed RLU. Any value that is under the three standard deviations is given a pass, while readings that are larger than three standard deviations fail.

In this lab, you will be evaluating the sensitivity of a rapid ATP system using purified ATP, pure bacterial cultures (*E. coli* K12), and food residues. This is based on a paper published by Flowers et al.[33] You will deposit your samples directly on the swab in a small volume (10 µl). This will eliminate errors due to uneven removal of bacteria and residues from surfaces. We should be using at least three swabs per data point in order to calculate a standard deviation at each point. However, in order to keep costs to a minimum, we will be using one swab for each of our samples.

All groups will be using micropipetters for measuring volumes. These are expensive and can be tricky to use. Please ask your instructor if you have any questions about their use. One key is to match the pipetter to the volume to be measured and use the correct tips for each pipetter (Table 18.1). If you use a different brand of micropipetter, make sure you use the appropriate micropipetter and tips for the volumes needed.

TABLE 18.1

Tips for Use in Gilson Pipetman® Models

Pipetman	Volumes	Tip Color
P1000	200–1000	Blue
P200	20–200	Yellow
P20	2–20	Yellow
P10	0.5–10	Clear (pink box)

■ III. METHODS

For this lab, the class will be divided into three groups, but students are responsible for understanding all aspects of this laboratory.

Group 1: ATP Standard Curve

This group will prepare a standard curve using pure ATP. For this laboratory, we are using ATP standard (Sigma Chemical, #FL-AAS). It will be supplied to you as a 1000 fmol/μl (or a 1 pmol /μl) solution.

Procedure

Prepare Dilutions

1. While wearing gloves (to reduce any transfer of ATP from your hands), carefully pour out the number of sterile 1.5 ml microcentrifuge tubes from a beaker and close their tops.

2. Label the concentration (fmol/μl) on the tubes with a permanent marker.

3. Prepare the dilution blanks from sterile deionized distilled water using a Pipetman or other micropipetter. First, prepare dilution blanks, as shown in Figure 18.2, using the supplied filter and sterilized deionized distilled water.

4. Remove the ATP stock (1000 fmol/μl) from ice and dilute as shown in Figure 18.2. Be careful not to touch the inside of the lid or tubes with your gloves, and be sure you mix all dilutions thoroughly using a vortex mixer before transferring to the next tube.

Inoculate the Swabs

Make sure all the dilutions were prepared prior to inoculating the swabs:

1. Pull the swab out of the top chamber of the PocketSwab. Do not touch anything with this swab. Transfer 10 μl of the dilution onto the swab portion.

2. Place the swab back into the PocketSwab. Do not twist down yet. Label each swab with the correct concentration.

3. Prepare all eight points of the standard curve, plus prepare a ninth point with sterile water (0 fmol/μl).

Take Reading in Luminometer

Do the following steps quickly in front of the luminometer:

1. Twist the swab all the way through the second membrane into the reaction mixture.
2. Tap the tube one to two times to help mix.
3. Remove the swab from the clear tube.
4. Tap once more.
5. Place PocketSwab in the luminometer and read according to the manufacturers' instructions.
6. Record the count in RLU.
7. Prepare a table with ATP concentration and light (RLU) to use as a handout for the next lab session for other groups.

Group 2: Detection Limit of PocketSwabs and Luminator for *E. coli* K12

You will receive a tube of overnight (16 to 18 h) aerobically grown *E. coli* K12 in tryptic soy broth or other nutrient media. First, you will wash the cells to assure that you are not measuring residual ATP from the growth broth.

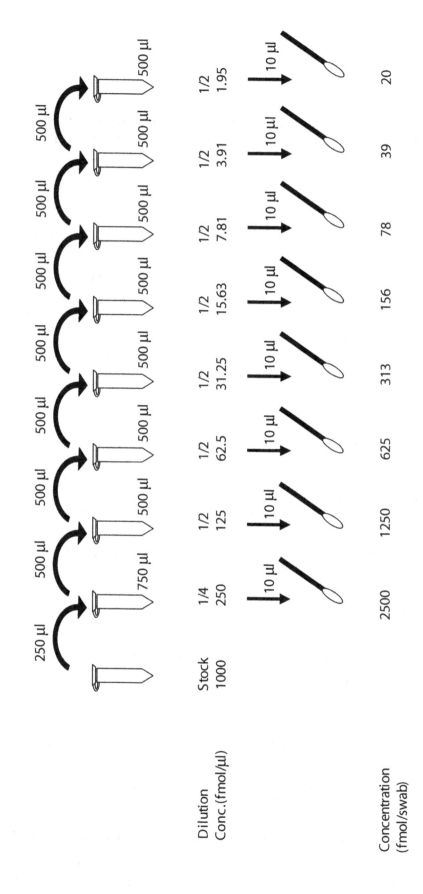

Figure 18.2 Dilution scheme for purified ATP detection curves.

Procedure

1. To wash cells, prepare two 1.5 ml microcentrifuge tubes with 1 ml overnight growth.
2. Spin down in microcentrifuge at 10,000 to 14,000 rpm for 5 min.
3. Carefully pour off used broth into a waste beaker. (This will be autoclaved later.)
4. Add 1 ml sterile phosphate buffered saline (PBS) to each tube, and suspend pellets by vortex mixing. If needed, use a micropipetter and gently agitate the pellet into solution by bringing the buffer in and out of the pipetter.
5. Repeat washing one more time (steps 1 through 4).
6. After the second wash, suspend each pellet in 1 ml PBS.
7. Combine both tubes into one blank sterile test tube. This is your 10^0 washed cell solution.
8. Prepare a series of 1/5 dilutions in 4 ml volumes of PBS, as outlined in Figure 18.3.

Inoculate the swabs according to Group 1, section B. Use 10 μl of each dilution for your Charm PocketSwab, and run an additional swab with 10 μl PBS for your negative control:

1. Perform ATP analysis according to Group 1, section C. Record RLU for each swab.
2. Determine cell numbers in your washed cell suspension by preparing 1/10 dilution series (to 10^{-6}). Spread plate (0.1 ml volume) final plated dilutions of 10^{-5} to 10^{-7} in duplicate on tryptic soy agar (TSA) or other nutrient media. Incubate at 37°C for 48 h. Next class period, count the colonies and calculate CFU/ml in the 10^0 washed cells.
3. Prepare a table with an RLU/swab, cell dilution, and cell number/dilution information to share with the rest of the class.

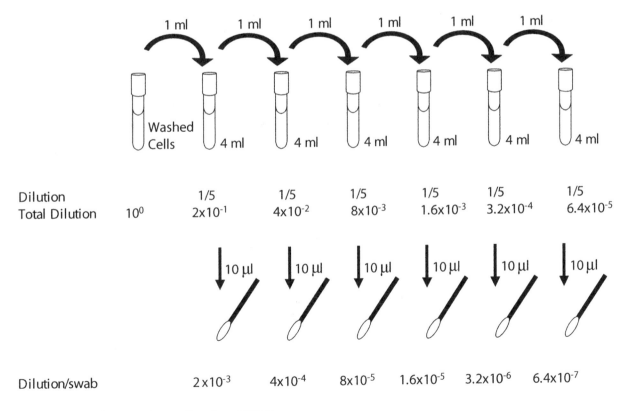

Figure 18.3 Dilution scheme for *E. coli* K12.

Group 3: Detection Limit of PocketSwabs and Luminometer for Detection of ATP in Food Residues

In this portion of the class, we will evaluate levels of ATP in two food residues: canned green bean liquid and fresh green bean rinse.

Procedure

1. Prepare fresh green bean rinse:

   a. Place 50 g of fresh green beans in a Stomacher bag.

   b. Add 50 ml of PBS buffer. Hand massage to create a fresh green bean rinse without damaging the green beans.

2. Canned green bean juice:

   a. Prior to opening the can of green beans, sanitize the can opener and the top of can with 70% ethanol.

   b. Open the can and directly sample juice for analysis.

3. Dilute both green bean juices using the dilution scheme shown in Figure 18.4. Note that we are performing a series of $1/2$ dilutions.

4. Inoculate the swabs according to Group 1, section B. Use 10 µl of each dilution for your Charm PocketSwab, and run an additional swab with 10 µl PBS for your negative control.

5. Perform ATP analysis according to Group 1, section C. Record RLU for each swab.

6. Prepare a table of RLUs to hand out at the next class session.

7. Prepare a second set of dilutions for your fresh green bean wash using standard $1/10$ dilutions (to 10^{-6}). Spread plate for a final dilution of 10^{-3} to 10^{-7} on TSA. Incubate at 35°C for 48 h.

8. During the next class period, count the plates, and calculate the CFU/ml on the fresh green bean wash. You can use this to approximate the amount of ATP measured due to bacteria vs. green bean residues.

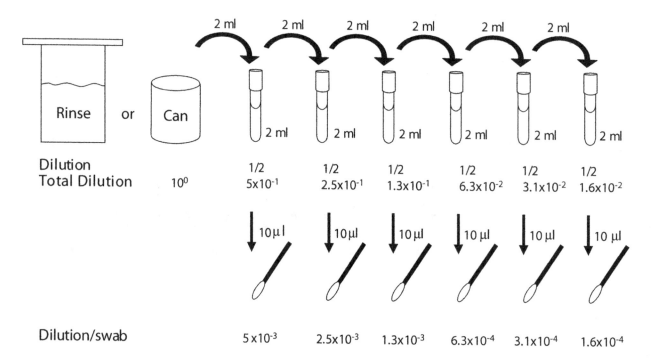

Figure 18.4   Dilution scheme for green bean juice.

■ IV. RESULTS

Using the data tables below, prepare the following plots:

Plot 1: ATP concentration (fmol/swab) vs. light (RLU/swab)
Plot 2: Bacterial cell concentration (log CFU/swab) vs. light (RLU/swab)
Plot 3: Canned bean juice (total dilution/swab) vs. light (RLU/swab)
Plot 4: Fresh bean wash (total dilution/swab) vs. light (RLU/swab)

ATP Concentration (fmol/swab) and RLU/Swab

ATP Concentration (fmol/swab)	RLU/Swab

E. coli K12 Count

Dilution as Plated	CFU/Plate	CFU/Plate	Average CFU/Plate
10^{-5}			
10^{-6}			
10^{-7}			

CFU/ml =

E. coli K12 CFU/ml, CFU/Swab, and RLU/Swab

E. coli K12 Dilution	RLU/Swab	CFU/Swab

Canned Green Bean Dilutions and RLU/Swab

Dilution of Canned Green Bean Juice	RLU/Swab

Fresh Green Bean SPC

Dilution as Plated	CFU/Plate	CFU/Plate	Average CFU/Plate
10^{-3}			
10^{-4}			
10^{-5}			
10^{-6}			
10^{-7}			

CFU/ml =

Fresh Green Bean Rinse Dilution, CFU/Swab and RLU/Swab

Bean Rinse Dilution	RLU/Swab	CFU/Swab

■ V. DISCUSSION QUESTIONS

1. Study Plot 1. Was there a linear relationship between the concentration of ATP and light? What was the lowest detectable concentration of ATP?

2. Study Plot 2. Was there a linear relationship between the amount of light and cell numbers? What was the lowest detectable concentration of bacterial cells your luciferase tests? Was this comparable to your results from Laboratory 17?

3. Study Plot 3. At which dilution was ATP no longer detected in each of the green bean juices?

4. Why did we not calculate the CFU/ml of the canned green bean juice? Canned green beans have long been dead, why do you think ATP was able to survive the thermal processing?

5. Identify some potential errors associated with our experimental procedure. What could you do to minimize these errors?

■ LABORATORY NOTES

■ LABORATORY NOTES

REFERENCES

1. Swanson, K.M.J., Busta, F.F., Peterson, E.H., and Johnson, M.G., Colony count methods, in *Compendium for the Microbiological Examination of Foods*, 3rd ed., Vanderzant, C. and Spittstoesser, D.F., Eds., American Public Health Association, Washington, D.C., 1992.

2. Mossel, D.A.A., Corry, J.E.L., Struijk, C.B., and Baird, R.M., *Essentials of the Microbiology of Foods: A Textbook for Advanced Studies*, John Wiley & Sons, New York, 1995.

3. ICMSF, *Microorganisms in Foods 2. Sampling for Microbiological Analysis: Principles and Specific Applications*, University of Toronto Press, Toronto, Canada, 1984.

4. Garthright, W.E., Most probable number from serial dilutions, in *Bacteriological Analytical Manual*, 8th ed., FDA, Ed., AOAC International, Arlington, VA, 1995. http://www.cfsan.fda.gov/~ebam/bam-toc.html.

5. Benson, H.J., *Microbiological Applications: A Laboratory Manual in General Microbiology*, 3rd ed., Wm. C. Brown Company Publishers, Dubuque, IA, 1984.

6. Gerhardt, P., Murray, R.G.E., Wood, W.A., and Krieg, N.R., Eds., *Methods for General and Molecular Bacteriology*, American Society for Microbiology, Washington, D.C., 1994.

7. Madigan, M.T., Martinko, J.M., and Parker, J., *Brock Biology of Microorganisms*, Prentice Hall, Upper Saddle River, New York, 2000.

8. Tournas, V., Stack, M.E., Mislivec, P.B., Koch, H.A., and Bandler, R., Yeasts, molds and mycotoxins, in *Bacteriological Analytical Manual*, 8th ed., FDA, Ed., AOAC, Gaithersburg, MD, 1998.

9. Mislivec, P.B., Beuchat, L.R., and Cousin, M.A., Yeasts and molds, in *Compendium of Methods for the Microbiological Examination of Foods*, 3rd ed., Vanderzant, C. and Spittstoesser, D.F., Eds., American Public Health Association, Washington, D.C., 1992.

10. Cray, W.C., Abbott, D.O., Beacorn, F.J., and Benson, S.T., Detection, isolation and identification of *Escherichia coli* O157:H7 and O157:NM (non-motile) from meat products, revision #2, 2-23-01, in *Microbiology Laboratory Guidebook*, 3rd ed., USDA, FSIS, OPHS, http://www.fsis.usda.gov/ophs/microlab/mlgbook.htm, 1998.

11. Andrews, W.H., June, G.A., Sherrod, P.S., Hammack, T.S., and Amaguana, R.M., Salmonella, in *Bacteriological Analytical Manual*, 8th ed., FDA, Ed., AOAC, Gaithersburg, MD, 1995.

12. Holt, J.G., Krieg, N.R., Sneath, P.H.A., Staley, J.T., and Williams, S.T., *Bergey's Manual of Determinative Bacteriology*, 9th ed., Williams & Wilkins, Baltimore, 1994.

13. Doyle, M.P. and Cliver, D.O., Vibrio, in *Foodborne Diseases*, Cliver, D.O., Ed., Academic Press, New York, 1990.

14. Adams, M.R. and Moss, M.O., *Food Microbiology*, 2nd ed., The Royal Society of Chemistry, Cambridge, 2000.

15. Elliot, E.L., Kaysner, C.A., Jackson, L., and Tamplin, M.L., *Vibrio cholerae, V. parahaemolyticus, V. vulnificus* and other *Vibrio* spp., in *Bacteriological Analytical Manual*, 8th ed., FDA, Ed., AOAC, Gaithersburg, MD, 1995.

16. Stern, N.J., Patton, C.M., Doyle, M.P., Park, C.E., and McCardell, B.A., *Campylobacter*, in *Compendium of Methods for the Microbiological Examination of Foods*, 3rd ed., Vanderzant, C. and Spittstoesser, D.F., Eds., American Public Health Association, Washington, D.C., 1992.

17. Ransom, G.M. and Rose, B.E., Isolation, identification and enumeration of *Campylobacter jejuni/coli* from meat and poultry products, in *FDA/FSIS Microbiology Laboratory Guidebook*, 3rd ed., http://www.fsis.usda.gov/ophs/microlab/mlgbook.htm, 1998.

18. Jay, J.M., *Modern Food Microbiology*, 6th ed., Chapman & Hall, London; New York, 2000.

19. Pennie, R.A., Zunnino, J.N., Rose, E., and Guerrant, R.L., Economical, simple method for production of the gaseous environment required for cultivation of *Campylobacter jejuni*, *J. Clin. Microbiol.*, 20, 320–322, 1984.

20. Bennet, R.W. and Lancette, G.A., *Staphylococcus aureus*, in *Bacteriological Analytical Manual*, 8th ed., FDA, Ed., AOAC, Gaithersburg, MD, 1998.

21. Doyle, M.P., Beuchat, L.R., and Montville, T.M., Eds., *Food Microbiology: Fundamentals and Frontiers*, American Society for Microbiology, Washington, D.C., 1997.

22. Hitchins, A.D., *Listeria monocytogenes*, in *Bacteriological Analytical Manual*, 8th ed., FDA, Ed., AOAC, Gaithersburg, MD, 1998.

23. Isolation and identification of *Listeria monocytogenes* from red meat, poultry, egg and environmental samples, revision #3 4/29/02, in *USDA/FSIS Microbiology Laboratory Guidebook*, 3rd ed., http://www.fsis.usda.gov/ophs/microlab/mlgbook.htm, 1998.

24. Prescott, L.M., Harley, J.P., and Kein, D.A., *Microbiology*, 2nd ed., Wm. C. Brown Publishers, Dubuque, IA, 1993.

25. Peleg, M. and Cole, M.B., Reinterpretation of microbial survival curves, *Crit. Rev. Food Sci.*, 35, 353–380, 1998.

26. Dryer, J.M. and Deibel, K.E., Canned foods—test for commercial sterility, in *Compendium of Methods for the Microbiological Examination of Foods*, 3rd ed., Vanderzant, C. and Spittstoesser, D.F., Eds., American Public Health Association, Washington, D.C., 1992.

27. Labbe, R.G., Recovery of spores of *Bacillus stearothermophilius* from thermal injury, *J. Appl. Bacteriol.*, 47, 457–462, 1979.

28. Vareltzis, K., Buck, E.M., and Labbe, R.G., Effectiveness of a betalains/potassium sorbate system versus sodium nitrate for color development and control of total aerobes, *Clostridium perfringens* and *Clostridium sporogenes* in chicken frankfurters, *J. Food Prot.*, 47, 532–636, 1984.

29. Landry, W.L., Schwab, A.H., and Lancette, G.A., Examination of canned foods, in *Bacteriological Analytical Manual*, 8th ed., FDA, Ed., AOAC International, Gaithersburg, MD, 1998.

30. Thomas, L.V. and Wimpenny, J.W.T., Investigation of the effect of combined variations in temperature, pH, and NaCl concentration on Nisin inhibition of *Listeria monocytogenes* and *Staphylococcus aureus*, *Appl. Environ. Microbiol.*, 62, 2006–2012, 1996.

31. Marriot, N.G., *Principles of Food Sanitation*, 4th ed., Aspen Publishers, Gaithersburg, MD, 1999.

32. Banwart, G.J., *Basic Food Microbiology*, 2nd ed., Chapman & Hall, London; New York, 1989.

33. Flowers, R., Milo, L., Myers, E., and Curiale, M.S., An evaluation of five ATP bioluminescense systems, *Food Qual.*, 23–33, 1997.

INDEX